ROBERT PAUL SHAY, JR.

British Rearmament in the Thirties

Politics and Profits

D0142016

PRINCETON UNIVERSITY PRESS
PRINCETON, NEW JERSEY

Published by Princeton University Press, Princeton, New Jersey
In the United Kingdom: Princeton University Press
Guildford, Surrey

Library of Congress Cataloging in Publication Data
will be found on the last printed page of this book

Publication of this book has been aided by a grant
from The Andrew W. Mellon Foundation

Printed in the United States of America
by Princeton University Press
Princeton, New Jersey

FOR MISSY

CONTENTS

CONTENTS

ACKNOWLEDGMENTS

LIKE all books, this one could never have been written without the kindness and cooperation of a great many individuals and institutions. While I have the space to name but a few, I extend my sincerest thanks to them all.

By their helpfulness, the men and women of the Public Record Office made my research in the government archives a true pleasure. That same trait characterizes the archivists at the Churchill College Library in Cambridge, and especially Mr. A. D. Childs who went out of his way to arrange for me to use the Weir papers while the library was under construction. The Institute for Historical Research kindly put their facilities at my disposal while I was in London. A Herbert Lehman Fellowship from New York State provided funds for my research.

For granting me permission to reproduce copyright material under their control I wish to thank the following: Viscount Caldecote for excerpts from his father's diary; Lord Hankey for his father's diary, letters, and papers; Viscount Weir for his grandfather's letters and papers; and above all the Controller of Her Majesty's Stationery Office for the vast amount of material in the archives of the Public Record Office that forms the basis of this book. The cartoons by David Low are reprinted by arrangement with the Trustees and the *London Evening Standard*. The Strube cartoons are reprinted with the permission of the *London Evening Standard*.

I am especially indebted to Professor Herman Ausubel, who encouraged me to study modern British history and sponsored the dissertation upon which this book is based. Professors Harold Barger, Robert Paxton, Michael Edelstein, Neville Thompson, Keith Middlemas, and Donald Lammers all read and provided useful criticism of the book while it was in draft form. My father, Robert P. Shay, Sr., gave me a number of helpful recommendations for revisions as well as some much needed assistance in clarifying my often murky economic prose. To Professor Arno Mayer I owe a special thanks not only for his thoughtful suggestions for strengthening this study, but, more importantly, for provoking my interest in the questions that brought me to it. The ongoing discussions about history and the historian's craft with my friends Larry Abrams and Bob Sonenthal have made this a far better book than it otherwise would have been. Of course the responsibility for any shortcomings in its style, content, or interpretations is mine alone.

I would be remiss if I did not acknowledge the prompt and helpful assistance I have received from Joanna Hitchcock and Carol Orr at the Princeton University Press in preparing this manuscript for publication.

Finally, the profoundest thanks of all I owe to Missy, my wife, who has provided for us both while I have shuffled the papers of history.

New York City
March, 1976

ABBREVIATIONS

C.I.D.	Committee of Imperial Defence
C.O.S.	Chiefs of Staff Sub-Committee of the C.I.D.
D.C.M. (32)	Ministerial Disarmament Committee
D.P. (P)	Defence Plans (Policy) Committee
D.P.R. (D.R.)	Defence Policy and Requirements Committee
D.R.C.	Defence Requirements Sub-Committee of the C.I.D.
F.B.I.	Federation of British Industries
N.D.C.	National Defence Contribution
P.S.O.	Principal Supply Officers Sub-Committee of the C.I.D.
S.B.A.C.	Society of British Aircraft Constructors
T.A.	Territorial Army

BRITISH REARMAMENT
IN THE THIRTIES

Needs must when the Devil drives.

*a northcountry
expression*

INTRODUCTION

$\mathbf{B}$RITAIN'S rearmament in the 1930s was the largest, most expensive program of any kind ever undertaken by a British government in time of peace. Initiated in response to the Japanese threat to the Empire, the program was spurred to full stride by the menace posed to the home islands by the burgeoning German air force. Although the British program was unprecedented in scope and size, it failed to keep pace with what Lord Weir, one of the Government's advisers, referred to as the "German Fantasy." Economic and political considerations prevented Britain from pursuing rearmament on the German scale. By 1938 Britain had reached the decision that she could not afford to rearm to the degree necessary to meet Germany on equal terms. The policy of appeasement was evolved in the belief that it might compensate for Britain's military weakness, but within months of the signing of the Munich agreement it became clear that the hope was false. By September 1939 Britain found herself involved in a war she was inadequately prepared to fight.

The purpose of this study is to examine the considerations that shaped Britain's rearmament policy, concentrating especially on those that led the Government to decide that rearmament had to be limited. While it is clear that the external threats to the security of the Empire and the nation made some form of rearmament imperative, how and why the Government chose to impose restraints on that rearmament are less apparent. At the heart of the decision was a complex combination of economic and political considerations that pertained only indirectly to Britain's military security. The priorities accorded those considerations directly reflect the Government's most fundamental assumptions about the nature of British society, and reveal much about its concerns for the future of British society.

Implemented over the strong objections of the armed services and the Foreign Office, the restraints affected the financing of rearmament and the Government's ability to organize it, thus slowing the rate at which it progressed. The financial restraints limited the level of expenditure on arms, while the organizational restraints, which prohibited intervention in the private sector of the economy, limited the Government's ability to increase arms production, and made it virtually impossible to insure that the resources available for defence were put to optimal use.

The process by which rearmament policy was formulated provides the focus for this study. Centering on the politics of decision making within the Government, that process encompassed an array of disparate political activities. On the cabinet level the complexities of interdepartmental politics, the influence of advisers, and the political strength of individual members of the Cabinet combined to form the setting in which decisions were made. Political struggles within the party that controlled the Government and the interests of the constituencies represented by that party often had a direct effect on those decisions. Although parliamentary politics obliged the Government to publicly explain and defend its policies, the impact those politics had on rearmament was limited due to the complexity of the issues involved and the size of the Government's majority.

Because of its complexity and magnitude the rearmament program affected and was affected by a multiplicity of issues, all of which the Cabinet had to consider in the course of formulating Britain's defence policy. Consequently, to understand that policy it is necessary to deal with questions ranging from global military strategy and labor relations to fiscal policy, and to evaluate the significance of the decisions made on each in terms of rearmament. To accomplish this an effort must be made to determine not only the positions of individuals, departments, and other interested parties on each question, but to ascertain the reasons underlying those positions. The purpose of this approach is to study rearmament from the point of view of those who formulated the policies that governed it by reconstructing the reality of the historical situation as they perceived it.

The sources on which this work is based were chosen with that purpose in mind. Among those sources the government papers in the Public Record Office in London were by far the most important, providing the preponderance of the material used in this study. Private papers of individuals involved in rearmament supplemented that material, while information published in the press and the proceedings of Parliament provided an insight into the public's perspective on the subject. Although the secondary literature on Britain in the thirties deals only tangentially with rearmament, it was helpful in establishing the political and diplomatic context in which the rearmament program evolved.

Of the papers at the Public Record Office those of the Cabinet, the Committee of Imperial Defence, and their respective subsidiary committees were important in indicating the way in which rearmament policy developed. They set forth the important issues and indicated the positions of the various parties on them. The Treasury papers provided detailed analysis of issues concerned with rearmament, and revealed the considerations underlying the Treasury's positions on those issues. Because the Treasury dominated the development of Britain's defence policy during this period, its papers are the most important in understanding the policy that emerged. The papers of the Admiralty, Army, and Foreign Office were not consulted to a great extent as their positions on policy were clearly presented in the Cabinet and Committee of Imperial Defence records. The files of the Air Ministry were consulted with respect to contracting and its relationship with the aircraft industry, but were not as complete as those of the Treasury.

The most useful of the private papers pertaining to rearmament were in the archives of the Churchill College Library in Cambridge. Of those the Weir papers were both the most complete with respect to rearmament and the most important. Also at Churchill College were the Hankey papers, which deal extensively with a wide range of topics, the Swinton papers, which proved disappointingly sparse for the period during which he was Secretary of State for Air, and excerpts from Thomas Inskip's diary concerning the Munich crisis and the period shortly thereafter. Though interesting in some respects, the papers of Templewood (Hoare) and Baldwin at the Cambridge University Library and of Runciman at the University of Newcastle-upon-Tyne yielded little useful information concerning rearmament.

5

For various reasons the papers of several important individuals in the National Government including Chamberlain, Simon, and Horace Wilson were not available for study.

A number of British trade associations including the Association of British Chambers of Commerce, the British Steel Corporation, and the Society of British Automobile Manufacturers were kind enough to allow access to their archives, however their contents proved to be of little relevance to this study. The Society of British Aerospace Companies, whose predecessors had run afoul of the government over profiteering, on the other hand, were demonstrably uncooperative in response to a request for access to their files.

Although the primary sources that were available for this study left some questions of detail unanswered, they were remarkable for the breadth and depth of enquiry that they allowed. They made it possible to ascertain what the central issues concerning rearmament were, to evaluate the positions of the interested parties on those issues, and to determine how and why decisions were reached. In view of the complexity and importance of the rearmament question, it is difficult to conceive of a more complete and useful record being available.

The secondary literature pertaining to rearmament was less useful. Essentially that literature falls into four categories: official civil histories of the Second World War, diplomatic histories, political biographies, and monographs on government and policy making. Within the literature there is a marked distinction between the books whose authors had access to the government documents for the period and those whose authors did not. At the heart of that distinction is the sureness and accuracy with which the policy-making process of the government is discussed.

The authors of the civil histories of the Second World War were the first to have access to those documents. Commissioned to do their work shortly after the war had ended, they produced a remarkable series of studies of the management and administration of the war economy, a series that remains an invaluable source of information and direction for anyone studying wartime Britain. Until 1967, when the government shortened the duration of the closure of the official papers from fifty to thirty years, those his-

tories constituted the main source of information on British rearmament. Scholars who have since had the opportunity to use the
official documents fully appreciate how well the authors of the
official histories did their work.

Despite their high quality, however, the civil histories are less
than satisfying for anyone seeking to develop a coherent concep-
tion of Britain's efforts to rearm. Because they focus primarily on
the period after the outbreak of the war, the accounts they provide
of the prewar armament effort are necessarily cursory and frag-
mented. The format of the series, which was established by the
government, also presents problems as there are no citations of
the documents on which the histories are based. Moreover, the
civil servants who often played such important roles in the
decision-making process are referred to anonymously. Despite
these shortcomings W. K. Hancock and M. M. Gowing's *The
British War Economy*, M. M. Postan's *British War Production*,
and especially William Ashworth's *Contracts and Finance* are of
great value in clarifying the issues and establishing the dimensions of the rearmament problem.

During the period after the war, historians who wrote about Britain in the thirties focused their attention primarily on foreign policy, and especially on the question of appeasement and Munich.
Using the published diplomatic documents on the period along
with material presented in evidence at the Nuremberg trials, John
Wheeler-Bennett in 1948 wrote his superb *Munich: Prologue to
Tragedy*, which set the standard for work on the subject. While
much more was subsequently written about Munich, appeasement, and the men responsible, it was not until 1968, when Keith
Robbins published *Munich, 1938*, that Wheeler-Bennett's work
was substantially improved upon. Both authors give detailed and
convincing accounts of the diplomatic maneuverings of the parties
involved in the crisis, but Robbins takes into greater account the
role played by domestic political considerations in its development. Neither, however, recognized the connection between the
economic restraints that had been placed on British rearmament
and the policy of appeasement that the Government chose to pursue.

It was left to Keith Middlemas to make that connection in his

Diplomacy of Illusion. Published in 1972, it was the first major study of British foreign policy in the late thirties to be based on the newly opened official government documents. Although he acknowledges the relationship between the economic considerations that caused the Government to limit rearmament and Britain's foreign policy, Middlemas accords them only secondary importance in his analysis of that policy's failure, choosing instead to attribute the failure to the inadequacy of Britain's political leadership.

In doing so he carries on the great liberal historiographical tradition that holds the leaders of a nation accountable for its fate. By thus focusing their criticism on a few powerful individuals, historians in this tradition are able to avoid the often disquieting insights that may emerge when the functioning of the political and economic structure in which those individuals operated is critically examined. Such insights can raise fundamental questions about the sources of power and influence and the interests they serve in not only the society the historian is studying but in his own society as well.

The strength of this tradition among British historians is underscored by their passion for studying their national history in the context of the lives of their political leaders. It is therefore not surprising that biographies dominate the literature on British politics in the thirties. In the study of British rearmament the primary value of political biographies lies in the information they provide about the political maneuverings, usually played out in private, that could and sometimes did affect the formulation of policy. Their value as sources of information about the development of the rearmament program, however, is limited by their form. Because the subjects of such biographies are public figures who in the course of their careers became involved in a wide range of important issues, their biographers are constrained by space and time from exploring any of those issues beyond the extent of their subject's involvement with them. Consequently complex issues like rearmament are usually discussed from a rather narrow perspective.

This being said, there are four biographies that discuss rearmament at some length and provide useful information concerning their subject's participation in the rearmament effort. Robert Rhodes James' *Churchill: A Study in Failure* devotes a chapter to Churchill's ongoing and often trenchant criticism of the Government's program, while Keith Middlemas and John Barnes spend considerable time in their biography of Baldwin detailing and sometimes overstating his role in getting the rearmament effort under way. W. J. Reader's biography of the Government's leading civilian adviser on rearmament, Lord Weir, provides an extensive account of the effort to convert segments of British industry to the production of war materiel, especially aircraft, and is the most useful single work on the subject. The third volume of Stephen Roskill's *Hankey: Man of Secrets, 1931–1963* is invaluable for the insights it provides into the politics of policy making under the National Government as well as for the new information it reveals about Hankey's contribution to the evolution of British rearmament.

Because of the emphasis placed on political biography, there are few monographs on British politics in the thirties. D. C. Watt's *Personalities and Policies*, which appeared in 1965, is in fact a collection of essays that raised for the first time the issue of the role of influential civil servants like Maurice Hankey and Warren Fisher in the formulation of policy. While rearmament is not dealt with in any of the essays, Watt's book is important because it draws attention to the internal politics of the National Government. The only true monograph that addresses itself to an aspect of the rearmament question is *Decision By Default*, Peter Dennis's study of the Government's ongoing opposition to the introduction of conscription during peacetime. Focusing on the debate in the Cabinet over the role of the Army in the next war, and the part the issue of conscription played in it, Dennis's book provides some useful information on the Army's vicissitudes under the National Government.

Although the secondary literature on Britain in the thirties is extensive, it fails to come to grips with the one issue that dominated the considerations of the Government throughout the balance of the decade, rearmament. Concentrating as it does on personalities and diplomacy, that literature seldom delves into the complex relationship between economics and politics that lies at the heart of the political process.

It is into that relationship that this study inquires. Because rearmament was vital to Britain's security and had significant ramifications for the national economy, the manner in which it was to be carried out was a major source of interest and concern to many of the most powerful and influential elements in Britain, both within government and without. Consequently the study of the way in which the course of British rearmament was determined provides an opportunity to observe the operation of the political process in its full scope. Such a study in depth makes it possible to identify and evaluate the fundamental concerns and considerations that motivated the National Government to pursue the policy it did.

CHAPTER I

The Coming of the National Government and the Pressures to Rearm

LABOUR AND THE FINANCIAL CRISIS

THE Cabinet's decision to revoke the Ten Year Rule[1] on March 23, 1932, in response to Japan's increasing threat to the Empire in the Far East, marked Britain's first halting step towards rearmament. The study of the process by which that decision was reached reveals many of the lines of division within the Government concerning defence policy, lines of division that were later to be brought into relief when the resurrection of Germany as a military force compelled the Government to make hard choices. Those choices, which determined the direction that British rearmament took, were influenced by many factors, both international and domestic, but none were more basic than the attitudes and predispositions of the men who made them. To gain an insight into those attitudes and predispositions it is useful to consider briefly the course of events that brought those men to power.

The event that had the most profound and pervasive impact on not only British rearmament but the whole course of history in the thirties was the Great Depression that ushered that decade in. An outgrowth of the economic malaise that gripped the capitalist world in the wake of the First World War, the depression was accelerated to the crisis point by the economic disasters that convulsed the American economy in 1929.[2] In Britain it caused the premature demise of Ramsay MacDonald's newly elected Labour Government enabling the Conservatives to regain the reins of power and retain them for the balance of the decade.

[1] The Ten Year Rule was the title given the decision that the armed services should assume that Britain would not become involved in a major war for ten years when they were drawing up their yearly budgets. See pages 11-24 for a more detailed discussion.

[2] John Kenneth Galbraith, *The Great Crash, 1929* (Boston: Houghton Mifflin, 1955) and Charles Kindleberger, *The World in Depression, 1929–1939* (London: Allen Lane, 1973).

11

When the Labour Government took office in June 1929, five months before the great crash of the American stock market, there were 1,164,000 unemployed insured workers in Britain, about 10 percent of the workforce. By July of the following year, after the economic effects of the crash had begun to take hold, 906,000 more people had joined their ranks, and thousands more were losing their jobs every week.[3] In the face of the economic forces at work there was little that the Government could do except to try to feed the unfortunate. Such efforts, however, combined with the rapid shrinking of revenue that resulted from the contraction of the economy produced a huge budget deficit that the Government had to cover by borrowing. Try as he did, the Chancellor of the Exchequer, Philip Snowden, could not bring the budget into balance as financial orthodoxy, his Treasury advisers, and the financial community told him he must if the nation's economy was to recover. By the spring of 1931 the business community[4] had begun a campaign to move the Government to adopt measures they believed would return the nation to the paths of fiscal respectability, and the economy to prosperity. Arguing that the Government was in effect the nation's largest employer, they contended that it should act, as any responsible business would when faced with a deficit, by reducing costs. They cited unemployment benefits and the dole as two areas where expenses might be reduced, and suggested cuts in the salaries of those in government employ such as teachers, civil servants, and servicemen as another measure by which economies might be affected. In the eyes of the business community the nation would only recover from the depression when industry recovered, and industry could only recover when the Government stopped squandering precious capital on nonproductive pursuits such as maintaining the standard of living of the unemployed, permitting it instead to be used to restore the productivity of the nation.[5] In July a parliamentary committee, the May Committee, dominated by Conservatives, is-

[3] Charles Loch Mowat, *Britain Between the Wars: 1918–1940* (Boston: Beacon Press, 1971), pp. 351–357.

[4] The term "business community" denotes a combination of the financial and industrial communities.

[5] The minutes of a secret meeting between the National Confederation of Employers Organisations and the Labour Government not only set forth the demands

sued a report recommending that the Government take the steps suggested by the business community, and warned that if such action was not taken the budget deficit would reach £120 million by the following April. The Labour Party found the recommendations, which placed the burden for recovery on those who had already paid the highest price, that is the poor and the unemployed, both politically and morally unacceptable, but could agree on no alternative. Meanwhile, by July 1931, unemployment had reached 2,800,000.[6]

At the same time that the May Committee was issuing its report, a second blow was striking the British economy. The international financial crisis that had been ravaging the German and Austrian banking systems spread to England. International confidence in the stability of the pound waned, and by the end of July the Bank of England had begun to lose gold at the rate of £2½ million a day. A £25 million credit on August 1 from the Federal Reserve Bank and the Bank of France did little to stem the tide, and by mid-August Bank officials were asking the French and Americans for an additional £80 million credit. Their response was that additional credits could not be extended until the Government took steps along the lines of those recommended in the May report to eliminate the budget deficit.[7] The alternative to obtaining the credits was to go off the gold standard, a measure that was considered heresy by the financial community, which saw it as an admission of national bankruptcy, totally unacceptable to the Bank of England, unlikely to receive the necessary Liberal support in Parliament, and anathema to Philip Snowden. Snow-

---

that the business community were pressing on the Government, but reveal the adversary relationship that existed between the two, a relationship very different from that which developed between the subsequent National Government and the business community. Illustrative of this is the fact that the President of the Employers Confederation, Lord Weir, became the National Government's leading adviser on rearmament. Cambridge, England, Churchill College Libary, Weir 12/1, 2/19/31, minutes of a meeting between the National Confederation of Employers Organisations and the Prime Minister and other members of the Government on the nation's economic situation.

[6] Mowat, *Britain*, p. 379. Noreen Branson and Margot Heineman provide a vivid picture of the impact on British society of the depression and the "recovery" that followed in their *Britain in the 1930's* (New York: Praeger Publishers Inc., 1971).

[7] Mowat, *Britain*, pp. 382–385.

den did his best to contrive a budget that would be acceptable to both his party and the international bankers, but failed, irreconcilably splitting the Labour Party in the process. He and MacDonald headed a small faction of the Labour Party that believed the national interest dictated that the budget deficit be reduced despite the effect that reduction would have on the working class.[8] They turned to the Liberal and Conservative parties to create a "National Government" to rescue the nation from the economic crisis. Formed on August 24, 1931, this Government, which was led by MacDonald but was wholly dependent on Conservative voting support in Parliament, quickly formulated a budget that met with the approval of the French and Americans who extended the £80 million credit. The sum was quickly consumed, and on September 21 Britain left the gold standard. Although orthodox finance had failed to save the gold standard, it had returned the Conservative Party to power in the guise of the National Government.

Parliament was dissolved on October 7, and an election was called for the 27th. In a short, vicious campaign, during which Snowden characterized his former party's program as "bolshevism run mad," while the Conservatives made dire predictions about the fate of the nation if Labour were returned, MacDonald asked that the National Government be granted a "doctor's mandate" to cure the nation's ills. This they were given as they were returned with 556 seats, 472 of which were held by Conservatives. Labour returned with only 46 members as compared to 287 in 1929. The depression had broken the back of the Labour Party. It was to remain a negligible political force throughout the thirties.[9]

THE NATIONAL GOVERNMENT

Although MacDonald, as Prime Minister in the new government,

[8] Robert Skidelsky's *Politicians and the Slump: The Labour Government of 1929–1931* (London: Macmillan, 1967) is the most interesting and useful account of the fate that befell MacDonald's government. He contends that the actual vote in the Cabinet on whether the Government should support Snowden's program of economies was 12-9 in favor of doing so, but that feeling on the issue was so strong that the Government could not carry on (see Skidelsky, p. 382).

[9] Mowat, *Britain*, pp. 411–412.

was its nominal head, the real power lay with the Conservatives on whose numbers the Government's majority in Commons was based. Their leader was Stanley Baldwin, who had come to leadership of the party in 1923, and had headed two Conservative governments in the twenties. An adept politician, his mastery of the arts of compromise and conciliation has led critics to question his decisiveness and his ability to lead. Justified as those criticisms may be, he held his party together when stronger leaders would only have torn it apart. In the National Government he held the position of Lord President, which allowed him to deal with problems as they arose. Neville Chamberlain, the new Chancellor of the Exchequer, was the number two man in the Conservative Party. A member of one of Britain's most distinguished political families, Chamberlain was Baldwin's opposite in almost every sense. His strengths were Baldwin's weaknesses, and his weaknesses Baldwin's strengths. Where Baldwin was conciliatory and politic, Chamberlain was rigid and abrasive; where Baldwin was sometimes indolent, indecisive, and short-sighted, Chamberlain was industrious, resolute, and insightful. The Treasury was the ideal place for Chamberlain. The department with the final say on all issues relating to finance, which is to say on all issues of substance, it required a man with the capacity to master a great deal of detail on a wide variety of subjects, and the ability to use that mastery as the basis for his decisions. In his five years as Chancellor of the Exchequer, Chamberlain, through the combination of his abilities with the powers of his office, became the most important decision maker in the Cabinet. Where Baldwin was a master politician, Chamberlain was a master administrator. Together they were a formidable pair.[10]

For all the rhetoric of the campaign, the National Government had little in the way of a program to bring about economic recovery. The low interest rates, a result of the nation's having left

[10] The best biography of Baldwin is Keith Middlemas and John Barnes, *Baldwin: A Biography* (London: Weidenfield and Nicolson, 1969). The best on Chamberlain remains Keith Feiling, *The Life of Neville Chamberlain* (Hamden, Connecticut: Arcon Books, 1970), a reprint of the original (London: Lowe and Brydone Ltd., 1946). At least two new biographies based on the Chamberlain papers and the government papers now available in the Public Record Office are presently under way.

the gold standard, enabled the Government to refinance the national debt at great savings to itself. Neville Chamberlain oversaw the institution of a system of tariffs, thereby bringing to pass a program advocated and bitterly fought for by his father before the First World War, but as trade was at a virtual standstill, it was of little economic consequence. The budget was balanced in the manner we have discussed above, and prices and wages were stabilized, all of which served to create a climate of confidence in the business community.

THE ECONOMY IN THE THIRTIES

Although the National Government did nothing in strictly economic terms that should have spurred the nation to recovery from the depression—the balancing of the budget should in fact have had just the opposite effect—the nation did recover. By 1935 the economy was beginning to prosper. However, because the new prosperity, based on and stimulated by domestic consumption rather than trade, was of a different nature from the traditional British prosperity, the economic specialists of the time were slow in realizing in its early stages that recovery was occurring. The traditional indices by which prosperity was measured, trade and employment, never approached the pre-Great War levels. Trade remained stagnant because the rest of the world was not so fortunate as Britain in its economic efforts to shake off the depression, and unemployment never dropped below 10 percent throughout the thirties as great numbers of men who had earned their livelihoods in the once great export industries such as textiles and coal never again found work. The recovery was led by new consumer industries like those involved in building automobiles, electrical goods, and homes. Based close to their markets in the South and Southeast, they gave rise to a new, highly skilled workforce. The result was that the new prosperity was very regional in character, centered in the urban areas of the South and scattered through parts of the Midlands, while in other parts of the Midlands, Wales, and through much of the North, pockets of great poverty remained. There were towns in which as much as 80 percent of the population was unemployed throughout the thirties. Although these "depressed areas" were of concern to the Government, they were so removed both economically and physically from the centers of recovery that they tended to be forgotten as the nation moved on with its new business. Recovery was well under way in the eyes of business because the economy once again offered ample scope for remunerative returns on investment.[11]

The importance of the way the National Government came to power and the way it administered that power from the outset lies in what it reveals about the relationship between the business community and the Conservative Party, which was the dominant force within that Government.[12] Central to this relationship was the fact that the business community was the core of the party's political constituency. They shared the understanding that in an economy such as Britain's, the business of the nation was business, that the prosperity of the nation flowed from the prosperity of business, and therefore that a government that ruled in the interest of business ruled in the interest of the nation as well.[13] The Labour Government fell because, while it refused to recognize the validity of the identity of interests set forth above, it lacked the power as well as, one suspects, the will to change the economic context so that that identity would no longer hold true. The Labour Party rightfully refused to sacrifice the interests of

[11] The two best works on the British economy in the thirties are: Derek Aldcroft, *The Inter-war Economy: Britain, 1919–1939* (New York: Columbia University Press, 1970) and H. W. Richardson, *Economic Recovery in Britain: 1932–1939* (London: Weidenfield and Nicolson, 1967).

[12] The relationship between interest groups and political parties is discussed by S. E. Finer in *The Anonymous Empire* (London: Pall Mall Press, 1958) and in "The Political Power of Private Capital," *Sociological Review*, New Series, Vol. 4, no. 1 (July 1956). Two contemporary studies that discuss this relationship with respect to the National Government's efforts to encourage Britain's two great stagnating industries, iron and cotton, to rationalize their production are: Ernest Davies, *"National" Capitalism: The Government's Record as Protector of Private Monopoly* (London: Victor Gollancz, 1939) and Arthur Lucas, *Industrial Reconstruction and the Control of Competition: The English Experiment* (New York: Longmans, Green and Co., 1937). The subject remains to be studied using corporate archives and government records, both of which are now available to scholars.

[13] Contributing materially to this shared understanding was the fact that a large percentage of Conservative MP's came from business backgrounds. In 1935, 181 of the 415 Conservative MP's sitting in the House of Commons held between them 775 company directorships. Simon Haxey, *Tory M.P.* (London: Victor Gollancz, 1939), p. 35.

17

its constituency, the working class, to those of the business community. In the economic context of the time, however, it had little choice but to step down because the short-term interests of its constituency did not coincide with the interests of the nation. It was to be fourteen years before it had another opportunity to alter that situation.

In the intervening period the relationship between business and the governing Conservative Party was to influence the preparation of the nation for war, as well as its eventual conduct of that war. It was a relationship that was to be subject to considerable stress as the interests of business and those of the nation came to diverge under war's threat, yet it was a relationship that persisted because both parties feared the consequences of the alternative.

THE INTERNATIONAL SITUATION, 1932

When the National Government took office in November after their overwhelming victory at the polls, it was clear that the depression had already begun to affect the international tranquility that had prevailed in the aftermath of the Great War. On September 18, 1931, Japan had invaded Manchuria in a quest to augment her limited supplies of natural resources and to create a new market for her manufactured goods. The preceding January Germany had set forth her demands that the Versailles limitations on the size of her armed forces be eliminated, and that she be allowed to create a force that would enable her to defend herself from her enemies. This demand too proceeded as much from the imperatives of economics as from those of nationalism.

The International Disarmament Conference scheduled to begin in Geneva in February 1932 offered the primary basis for hope that the thirties and the decades to follow would see a world at peace. Enthusiasm for a successful result was great, especially in Britain, where a great many people had come to the conclusion that disarmament offered the only sane assurance of world peace. For them the thirties were to be a confusing and bitterly disappointing time. In actuality the conference foundered on the irreconcilability of France's desire for eternal security from the threat of German attack and on Germany's desire to return as an equal to the world of nations, which of course meant to rearm to the level that other nations were armed. Germany left the conference in September 1932, only to return. In January 1933 Hitler was appointed Chancellor, and used the conference for his own purposes while he consolidated his internal political position before withdrawing for good in October of that year. The British worked hard to make the conference a success, although they too balked at proposals that would have affected their special concerns. They were the nation with the most to lose from the failure of the conference, for they were the nation with the most extensive international commitments and the least adequate military establishment to support them. The fate of the conference, however, lay in other hands, and was subject to other concerns.[14]

THE TEN YEAR RULE

Britain's rather palsied military stance was the result of thirteen years of untroubled international security, and the imperatives of budget balancing to maintain economic stability. In August 1919, when Britain was experiencing the economic instability that accompanied the massive and rapid demobilization that followed the First World War, the Cabinet formulated an assumption that was to serve as a guideline for the military services in their planning and budget making. The assumption was simply that there would be no major war for ten years from that date. This assumption, the Ten Year Rule, as it came to be known, lingered through the twenties, being reaffirmed at various times. In July 1928, Winston Churchill, then Chancellor of the Exchequer, suggested that the rule be altered so that the ten years in question would be moved forward each day the assumption was not altered, rather than counted from the last date the assumption was reaffirmed. Although the chiefs of staff were not happy about the proposal, it was confirmed by the Cabinet.

The rule itself was an embodiment of the realities of the international situation in combination with the vicissitudes of the domestic economy. The nation had no credible enemies after the Great War ended. As a result the Army, which had been badly deci-

---

[14] Charles C. Bright, "Britain's Search for Security 1930-1936: The Diplomacy of Naval Disarmament and Imperial Defence" (Unpublished Ph.D. dissertation, Yale University, 1970), pp. 93–98.

mated in the course of the war, was demobilized completely and put on a professional basis. The plant that had been built in the course of the war to supply it with arms and ammunition was returned to civilian use. The Air Force, which had become one of the most powerful in the world, was allowed to deteriorate to the point where by 1931 it ranked only fifth.[15] Even the Navy was allowed to decay. In 1929 they had only fifty-six serviceable cruisers rather than the seventy they saw as necessary, and thirty-four of those would become obsolescent within the next decade.[16] The decisions that allowed this situation to come to pass were entirely rational. America on the sea and France on the ground and in the air were the only two threats the military could point to in their bids for funds, and they were not credible. The domestic social situation was far more threatening and in need of attention. As a result social expenditure as a percentage of the budget greatly increased, while military spending decreased. When the depression struck, military expenditure was further reduced in the Government's efforts to balance the budget. Stores of supplies were not replaced, building programs were suspended, and the military establishment was allowed to drop below the minimal levels of readiness required under the precepts of the Ten Year Rule. Under the circumstances, these measures were justified.

The Japanese invasion of Manchuria, or more precisely their siege of Shanghai, gave the British military the opening they required to bring the Ten Year Rule into question. The initial invasion had not in fact caused the Government much concern, preoccupied as it was with the financial crisis. Britain had sent troops into Shanghai as recently as 1927 to protect the trading interests of her nationals and restore the rule of law where the feeble central Chinese government was incapable of upholding it. The Japanese were giving the same reasons for their move. In January, however, when the Japanese began to battle for Shanghai, the realization of the vulnerability to Japanese attack of British interests in the Far East was dramatically brought home.[17] Although the Shanghai situation was settled within two months through negotiation, it

[15] CAB 24/224, CP 296 (31), 11/26/31 memorandum from the S/S for Air, Lord Londonderry, to the Cabinet on the international air situation.

[16] Bright, "Britain's Search for Security," p. 23.

[17] Christopher Thorne, *The Limits of Foreign Policy. The West, the League and the Far Eastern Crisis of 1931–1933* (New York: G.P. Putnam's Sons, 1972).

marked the beginning of a prolonged period of tension between British and Japanese interests throughout the area. This tension underscored Britain's military vulnerability, bringing about the demise of the Ten Year Rule and the onset of the first stage of British rearmament.

The revocation of the Ten Year Rule, though it proved to be of little more than symbolic importance, provides a good illustration of the way the decision-making process operates under the British system of cabinet government. Within this system the cabinet, made up of the leading members of the ruling political party, or in the case of the National Government, parties, is the ultimate decision-making body. Most of its members are heads of government ministries, usually holding the title of secretary of state. The cabinet minister's primary contact with his ministry is through the permanent under-secretary of that ministry. He is a career civil servant who has risen through the ranks, generally within his own department, to reach his position. The role of the permanent under-secretary is to keep the secretary of state informed about the activity of the ministry, and especially about the issues concerning the ministry that he will have to explain and often battle for in the cabinet and in Parliament. The importance of the permanent under-secretary in the decision-making process should not be underestimated. With his long experience and consequent familiarity with the issues he cannot help but persuade the secretary of state of the correctness of the ministry point of view. In that cabinet ministers seldom stay at the same post for more than a couple of years, they are largely reliant on the advice of their secretaries. It is this that accounts to a large degree for the continuity of policy between governments and within them as well.

The role of the cabinet is to discuss and ratify decisions that have been made elsewhere. Issues of political importance are often aired in cabinet before a decision is made, but the actual decisions are arrived at in discussion among the interested parties. In the case of an irreconcilable conflict it is hierarchy, not a democratic vote, that usually decides. As we shall see the opinion of the Chancellor of the Exchequer counts for more than that of the three service ministers combined.[18]

[18] The two standard works on cabinet government in Britain are: William Ivor Jennings, *Cabinet Government* (Cambridge: Cambridge University Press, 1959) and John P. Mackintosh, *The British Cabinet* (London: Stevens Press, 1962).

Questions of military and foreign policy were usually discussed in the Committee of Imperial Defence. The C.I.D., as it was called, was chaired by the Prime Minister, and its membership included the ca'.'net ministers from the Treasury, the Foreign Office, and the three armed services along with their respective permanent under-secretaries, or in the case of the services their chiefs of staff. This was the forum in which the issue of the Ten Year Rule was raised and discussed.[19]

Attached to the C.I.D. were several subcommittees to which it delegated problems, and which raised issues for it to consider. The most important of these subcommittees was the Chiefs of Staff Sub-Committee, or C.O.S. Made up of the Chief of the Naval Staff, the Chief of the Air Staff, and the Chief of the Imperial General Staff, the role of this committee was to coordinate the three services. In reality little was ever accomplished here because the three services were bitter rivals competing for limited funds. As a result, issues that caused conflict, which is to say most issues of any importance, were glossed over and passed on to the C.I.D. for the political ministers to settle. The Ten Year Rule, however, was a question on which the C.O.S. could readily agree. All of the services suffered under it, and all of them had been waiting for an opportunity to persuade the Cabinet to revoke it. The Japanese move on Shanghai provided that opportunity.

In a statement to the C.I.D. dated March 17, 1932,[20] the C.O.S. cited the weaknesses revealed by the Manchurian crisis as the basis for their argument that the Ten Year Rule should be suspended. They pointed out that

> we possess only light naval forces in the Far East; the fuel supplies required for the passage of the Main Fleet to the East and for its mobility on arrival are in jeopardy; the bases at Singapore and Hong Kong, essential to the maintenance of a fleet of capital ships on arrival in the Far East are not in a defensible

[19] Two somewhat dated studies of the C.I.D. are: Franklyn Arthur Johnson, *Defense by Committee: The British Committee of Imperial Defense 1885–1959* (London & New York: Oxford University Press, 1960), Field-Marshal Lord Milne, "The Higher Organisation of National Defence," *The Nineteenth Century and After*, March 1936, pp. 305–318.

[20] CAB 24/229, CP 104 (32), also C.I.D. 1082–B, 3/17/32, report by the C.O.S. subcommittee on the Ten Year Rule and the Far East.

condition. The whole of our territory in the Far East as well as the coast line of India and the Dominions and our vast trade and shipping lies open to attack.

The weaknesses of the other two services in relation to comparable world powers were also dealt with, and the C.O.S. baldly stated that Britain's armed services were unequipped to meet any of her potential military obligations due to the effects of the Ten Year Rule. In an appeal to which they felt the new Government would be responsive they decried the sacrifices forced on them by the Labour Government, stating that:

> Amid the colossal expenditure on development and unemployment, on which the May Report on National Expenditure sheds so searching a light, we find it difficult to believe that the relatively modest sums required to correct the more glaring defects in the Imperial Defence could not have been found.

They then made their proposals: a. that the Ten Year Rule should be cancelled; b. that expenditures for "purely defensive commitments" in the Far East should be made, that is, the strengthening of the naval base at Singapore; c. decisions on the strengthening of the services should not be delayed until the disarmament conference was concluded. They warned that, "We cannot ignore the Writing on the Wall."

The Treasury reply to this impassioned plea by the C.O.S., submitted to the C.I.D. under Neville Chamberlain's signature, is a classic statement of its basic rationale for opposing the services' requests for funds.[21] It was a rationale that was to remain constant throughout the thirties. Pointing out that "the fact is that in present circumstances we are no more in a position financially and economically to engage in a major war in the Far East than we are militarily," the Treasury explained that the Ten Year Rule "is no more than a working hypothesis intended to relieve the C.O.S. from the responsibility of preparing against contingencies which the Government believe to be either remote or *beyond the financial capacity of the country to provide against*." After reminding the C.I.D. that the financial health of the nation was of paramount

---

[21] CAB 24/229, CP 105 (32), also C.I.D. 1087–B, 3/17/32, note by Neville Chamberlain on the Ten Year Rule and finance.

importance, and that all other considerations were subordinate to it, the statement concluded "that today financial and economic risks are by far the most serious and urgent the country has to face and that other risks must be run until the country has had time and opportunity to recuperate and our financial situation to improve." The financial situation eventually did improve, and the services were granted larger funds to remedy their deficiencies, but the rationale remained: finance first.

On March 22 the C.I.D. met to consider the issues raised by the two papers.[22] Bolton Eyres-Monsell, the First Lord of the Admiralty, speaking on behalf of the services, reiterated that the Ten Year Rule was hardly fair to the services responsible for Britain's and the Empire's defence, and went on to say that the country would be horrified if they learned that it was the basis of defence planning. MacDonald and John Simon, the Secretary of State for Foreign Affairs, both men with philosophical commitments to disarmament, reluctantly agreed with the logic of the C.O.S. case, and Chamberlain hastened to make it clear that his Treasury paper was not meant to challenge the recommendations of the C.O.S., but merely to remind the committee of the financial situation. The recommendations of the C.O.S. report were accepted, and they went to the Cabinet for approval the next day. There the significance of the Treasury paper, which was submitted to the Cabinet along with the C.O.S. report, became clear.[23] The Cabinet accepted the cancellation of the Ten Year Rule without dissent; however it went on to state that this did not justify immediate increases in defence spending, and that in view of the disarmament conference the whole spending question should be studied further. The services had won the right to plan, but not to spend. The story of rearmament in the thirties revolves around this second battle.

THE BUDGET

The spending battle was fought yearly between the services and

---

[22] CAB 2/5, 255 (4), 3/22/32. (CAB 2/- will denote C.I.D. minutes in future footnotes.)

[23] CAB 23/70, 19(32)2, 3/23/32. (CAB 23/- will denote Cabinet minutes in future footnotes.)

the Treasury, as well as among the services themselves, in the formulation of the budget. The skirmishing began in the October preceding the year for which the budget was submitted, when the services as well as all the other spending departments were asked to submit preliminary estimates of their costs and expenses for the following fiscal year. At the same time the Treasury was going over the early revenue returns for the current year, comparing them with their earlier projections of that revenue, and trying to get an indication of the level of revenue they could expect the following year. By December a general picture of the situation had emerged and the services were informed of the boundaries within which the Treasury hoped to see their final estimates fall. In January the services submitted their final estimates and the Treasury went over them in detail suggesting cuts that could be made. The estimates were almost always higher than the limits the Treasury set, which were in turn usually lower than the levels at which the Treasury expected to finally settle. There was always give and take, but the bargaining was in dead earnest. Frequently there were differences over policy and principle that had to go to the Cabinet for final resolution. As a result the Treasury had to develop policy positions on military and foreign affairs as well as on the economy and the domestic situation. Through its control of finance the Treasury became the central body for the development and coordination of policy.[24]

While the final defence estimates were being settled along with those of the other departments, and ratified by the Cabinet, usually by the end of February, the revenue department was drawing up tax and borrowing programs to cover their cost. The budget was finally presented to the Parliament in the middle of April, where it was discussed and gone over for about two and a half months until it was passed on the third reading and sent to the monarch for the royal assent. The changes made by Parliament were seldom substantial if the Government and the Treasury had done their work thoroughly.

The cancellation of the Ten Year Rule came too late in the year

[24] Two useful essays on Treasury control are: Lord Bridges, *The Treasury*, The New Whitehall Series #12 (London: Allen and Unwin, 1964) and Samuel H. Beer, *Treasury Control* (Oxford: Clarendon Press, 1957).

to have any effect on the military appropriations in the 1932 budget, which in fact cut military spending to its lowest level in the interwar period.[25] Despite the austerity of the 1932 budget, there turned out to be a deficit of £32 million, which did not dispose the Treasury to loosen the purse strings for the services in 1933 despite the cancellation of the Ten Year Rule.[26] The only additional defence spending the Treasury anticipated was for the strengthening of the port at Singapore, which the Cabinet had approved in principle in October of 1932 subject to Treasury authorization of specific expenditures.[27] The Navy had other plans, however, and battled the Treasury into February for additional monies to make up stores depleted as the result of economies in the preceding years, as well as for authorization of an expensive new ship construction program.[28] The Treasury held that the economic situation was still too serious to take chances with, and would not give in. Eyres-Monsell managed to have the issue raised in Cabinet, something he technically needed the permission of the Treasury to do, but which he did without, and argued that with the Ten Year Rule no longer in effect the Navy was unequipped to meet Britain's global naval responsibilities. He demanded that if the Cabinet could not see fit to authorize the Navy's request, it should give an assurance "that they realise and take responsibility for the continued unreadiness of the Navy." This was actually something of a rhetorical gesture on the Admiralty's part, as under the constitution the Cabinet automatically assumed that responsibility when it authorized the budget. After Chamberlain reiterated the Treasury view on the seriousness of the economic situation, the Cabinet agreed to take the responsibility for the Navy's unreadiness, while admonishing the defence ministers to raise their questions through normal channels in the future.[29] The

[25] A.J.P. Taylor, *English History: 1914-1945*, The Oxford History of England, Vol. XV, edited by Sir George Clark (Oxford: Clarendon Press, 1965), p. 364. For the details of Britain's defence expenditures in the thirties see Appendix.

[26] T 171/296, 11/3/32, note on projected financial situation from Hopkins to Chamberlain. T 171/309, 1/14/33, note from Hopkins to Fisher and Chamberlain on the progress of the budget.

[27] CAB 23/72, 50(32)9, 10/11/32.

[28] CAB 24/237, CP 25(33), 2/8/33, memo by Eyres-Monsell to the Cabinet on the naval estimates.

[29] CAB 23/75, 9(33)3, 2/15/33.

Army and the R.A.F. were content to let the Navy launch this trial balloon to see whether the end of the Ten Year Rule had brought any change in the direction of the Treasury breeze. All it brought was the comment from Warren Fisher, the Permanent Treasury Secretary known for his colorful forthright candor, that "the Navy . . . is an example of arrogant inefficiency."[30]

Any belief that the Ten Year Rule in itself was the cause of the restraints put on defence spending should have been dispelled by this episode. The rule was simply an expression of the restrictions the economic situation put on the military. Even if the rule had never come into existence, defence spending would have been set at the same level, just as it was after the rule was revoked. In revoking the rule the Cabinet acknowledged the worsening of the world situation without making a commitment to deal with it. It was a commitment that was to be determined as much if not more by economic considerations than by those of military or diplomatic strategy.

GERMANY IN 1933

Hitler's appointment as Chancellor and his National Socialist Party's subsequent consolidation of power was a source of considerable concern to the Foreign Office and the Government. As early as February of 1933 gloomy forecasts about the future of the disarmament conference and Germany's consequent rearmament were being made. An arms race between Germany on the one hand and Britain and France on the other was predicted, followed by Germany's going to war in the East with an eventual European war resulting, all within three to four years' time.[31] One should hasten to add that this was the Foreign Office's darkest vision, and it was presented with the intention of motivating the Government to make even greater efforts to save the failing disarmament conference. That made the assessment of the situation no less valid however, and the French reading of the situation was just as pessimistic, if not more so. In fact the possibility of a preventive war

[30] T 161/580/35164/34, 4/1/33, Fisher note on service estimates.
[31] CAB 24/239, CP 52(33), 2/28/33, Simon on the crisis in Europe; CAB 24/241, CP 129(33), 5/7/33, Vansittart on the foreign policy of the present German government.

27

to remove Hitler was being discussed in high places in Paris in May and June.[32] Germany was rearming in flagrant violation of the Versailles Treaty, and made little effort to conceal the fact.[33] When she left the disarmament conference for the last time in October, it was clear that Britain was going to have a defence problem on the Continent to go with the one she already had in the Far East.

In light of this unpleasant if not unexpected change in the European situation, the C.O.S. drew up a new assessment of the imperial defence situation. They cited Germany as a profound threat to British security within three to five years, and emphasized the extreme unpreparedness of the Army for any type of continental role. It was an unpreparedness that could not be remedied within a short time, because Britain was lacking not only in guns, ammunition, and men, but in the plant to produce the guns and ammunition as well. It was a deficiency that had to be dealt with right away if the nation were to be ready when the test came.[34] At the 261st meeting of the C.I.D., the first since the termination of the Ten Year Rule, the C.O.S. paper was discussed. Chamberlain suggested that in view of the European developments the Government had to order its defence priorities, and he suggested that serious efforts be made to get on better terms with Japan in order to eliminate one of Britain's serious problems. Simon pointed out that to do so was to run a grave risk, for Japan had ambitions to become a great power in the Far East and the leader of "the Yellow Peoples," which would jeopardize Britain's interests in the area. Although the issue was not elaborated at this point, it was to become the center of an ongoing policy dispute between the Treasury and the Foreign Office and Admiralty, who both saw the question in the same way. The major revelation of the meeting was Chamberlain's indication that "the financial situation was not quite so difficult," and that the balance between military and financial considerations had shifted somewhat to the former. Chamberlain suggested that the services should work out pro-

[32] CAB 24/242, CP 151(33), 6/6/33, the French view of the world situation as reported by the Ambassador to France, Lord Tyrrell.

[33] CAB 24/242, CP 184(33), 7/14/33, on German rearmament.

[34] CAB 24/244, CP 264(33), 11/10/33, Review of Imperial Defence Policy, presented earlier as C.O.S. 310, and to the C.I.D. as 1113-B.

1. "Signs of Returning Prosperity"—While the Government was only considering rearmament in 1933, David Low and other advocates of disarmament were claiming that the British arms industry was doing well selling weapons to small African, Asian, and South American countries. *Evening Standard*, October 9, 1933.

grams for making good their deficiencies among themselves, and MacDonald took this one step further by proposing that the C.O.S., the secretary of the C.I.D., and the secretaries of the Foreign Office and Treasury should prepare such a program for presentation to the Cabinet.[35] This was the birth of the Defence Requirements Committee, to be known henceforth as the D.R.C.

THE DEFENCE REQUIREMENTS COMMITTEE

It was the D.R.C. that formulated the plans that were the basis of Britain's rearmament. The committee itself was made up of the most powerful and influential civil servants in British government at the time. The chairman was Maurice Hankey, the Secretary to the Cabinet and the C.I.D., posts he had held continuously since the First World War. A man of enormous capability, energy, and experience, he held positions that allowed him to survey the breadth of government policy without the bias inevitable in a ministerial secretary. His opinions, based on his long experience and his broad perspective, were respected if not always acted upon. A master of the bureaucracy he had done so much to shape, he knew how to get things done.[36] Warren Fisher, the Permanent Secretary to the Treasury, was a close friend of Hankey's, although, as we shall see, often an opponent on questions of policy. Fisher was second only to Hankey in influence and experience. As the Treasury Secretary he dealt with as wide a variety of issues as Hankey for, as we have seen, the Treasury concerned itself with all matters involving expenditure, which in effect meant everything. Fisher was especially interested in foreign policy, and held strong views on the subject, which he set forth in a vivid, uncompromising style. On the D.R.C. he became the dominant force.[37] The Permanent Under-Secretary of State, Robert Vansittart, was an astute observer of the world scene with a gift for crys-

[35] CAB 2/6, 261(1), 11/9/33.

[36] Captain Stephen Roskill's three volume biography of Maurice Hankey, *Hankey, Man of Secrets* (London: Collins, 1974) is an exhaustive study of the life and times of Britain's most powerful civil servant in the twentieth century.

[37] There is no full length biography of Fisher. D. C. Watt's sketch in *Personalities and Policies* (London: Longmans, 1965) is good in a general sense although it tends to overrate Fisher's influence somewhat.

tallizing his insights in the felicitous phrase.[38] Perceptive as he was, however, his influence was limited because he headed a department that operated within the limits of policy made elsewhere. Although his reading of the world situation was accurate, this did not give him the power to set the policy to deal with it.[39] The military men on the committee were Admiral Sir Ernle Chatfield from the Navy, Sir Edward Ellington from the R.A.F., and Field-Marshal Sir Edward Montgomery-Massingberd from the Army. They set forth the views of their respective services on the world situation and the recommendations of the services on what was required to equip the nation to meet the demands of that situation.

The first meeting of the D.R.C. took place on November 14, 1933, and it was agreed in principle that the committee should concern itself primarily with the nation's military needs, leaving questions of political and economic feasibility up to the Cabinet.[40] The committee then turned to the question of the threats to Britain's security. Vansittart expressed the belief that Germany was the primary threat in the long term, although the Foreign Office had not yet officially committed itself to that position. The committee agreed with that assessment,[41] then came to grips with the question of how Japan should be dealt with in view of it. The Navy had been pressing hard for authorization to build the fleet up to seventy cruisers to enable them to meet the Japanese threat in the Far East. They stressed that the whole Empire in the East from India to New Zealand was endangered, and they were supported in this view by the Foreign Office. The Treasury, however, citing the economic crisis, had refused to go along. Now that the Treas-

---

[38] Vansittart is another whose prose makes the documents in the Public Record Office more enjoyable. Ian Colvin has written a biography, *Vansittart in Office* (London: Gollancz, 1965), which is not very helpful. More rewarding is his autobiography, *The Mist Procession* (London: Hutchinson, 1958), if taken with a grain of salt.

[39] The amount of personal influence civil servants like Hankey, Fisher, and Vansittart had over broad policy decisions depended in large part on their relationships with the politically powerful in the Cabinet. Such influence was directly proportional to the political patron's power. It endured only as long as the civil servant's advice remained compatible with his patron's views and the patron remained in office.

[40] CAB 16/109, 11/14/33, 1st meeting of the D.R.C.

[41] CAB 16/109, 12/4/33, 3rd meeting of the D.R.C.

ury had acknowledged that the financial situation had eased, the Admiralty was clamoring for authorization to begin its cruiser-building program. The Treasury did not concur.

The Treasury argument, as expressed by Fisher in the D.R.C., was that because the nation lacked the resources to engage two first-class powers simultaneously, it had to decide which was the more serious threat and concentrate its resources on meeting it while trying to accommodate the other. In this case Germany had been cited by the D.R.C. as the more serious threat, therefore Britain should concentrate the bulk of her efforts on neutralizing that threat while trying to come to terms with Japan.[42] To get on good terms with Japan, Fisher recognized that it was first necessary to win her respect. Suggesting that that could be accomplished by a show of resolve in the Far East, he recommended the rapid completion of the base at Singapore and the modernization of part of the fleet for operations there. These projects could be accomplished in a relatively short time at little cost compared to that of building the fleet up to seventy cruisers, and would impress the Japanese. At the same time, by negotiating with Japan, Britain would demonstrate her independence from the United States in the Pacific. Fisher believed that it was useless to depend on the United States to keep Japan under control because he felt that the United States was totally unreliable in international affairs, one of his more strongly held views.[43] This was Fisher's and the Treasury's position on how the world situation should be handled. Underlying it was the concern that the nation, despite the improved economic outlook, could not afford to build the fleet that the Navy said was necessary, especially if it were going to have to find funds to counter the German threat as well.

The D.R.C. was not terribly enthusiastic about Fisher's plan. Hankey and Vansittart were concerned about the fate of the Empire if the Japanese could not be accommodated, or worse, went back on assurances of good intent—behavior they believed the Japanese to be perfectly capable of judging from their recent activity. The Navy was not taken with the plan because it denied

[42] CAB 16/109, D.R.C. 6, 2/12/34, letter from Fisher to Hankey.
[43] CAB 16/109, D.R.C. 12, 1/30/34, Fisher to the C.O.S. on the need to get on terms with Japan.

them cruisers they wanted. Rather than fight Fisher and the other two services, who opposed the Navy's building program because it meant less funds for them, the admirals decided to let Fisher have his way in the D.R.C. They felt they could get their program before the Cabinet in more favorable circumstances by way of the committee on the upcoming Naval conference in which they and the Foreign Office drew up Britain's position for Cabinet approval.[44]

As a result the D.R.C.'s first report embodied the Fisher plan of "showing tooth" in the Far East in order to create a climate in which accommodation with Japan could be reached. To deal with the German threat, the D.R.C. proposed a five year program to remedy the deficiencies of the armed forces. Under this program the R.A.F. would be brought up to a full fifty-two squadrons, the strength recommended by the air staff and approved by the Cabinet in 1923 but, due to subsequent economies, never achieved. The Army was to be able to put four infantry divisions, one tank brigade, and one cavalry division on the Continent within the five year period. No decision was made on naval construction, and no increase in the size of the fleet was recommended, but the report did include provisions for modernization of the existing fleet, the building of new bases and fueling stations, and an increase in the air arm. In all the program was to cost £71,322,000 not including any naval building that might be done.[45]

The Cabinet first discussed the report on March 14, 1934, but got no further than debating the question of whether it was feasible to reach an accord with Japan without endangering the Empire. No decision was reached.[46] Five days later the Cabinet took up the report again, this time addressing the question of whether Germany should be made the leading potential enemy. The Secretary of State for India, Samuel Hoare, doubted the wisdom of doing so, to which John Simon replied that although she was not presently a threat, the direction Nazism was taking would make her one in the future. After MacDonald persisted in putting the

---

[44] Bright, "Britain's Search for Security," p. 161.
[45] CAB 24/247, CP 64(34), 3/5/34, also referred to as D.R.C. 14, the D.R.C.'s first report.
[46] CAB 23/78, 9(34)13, 3/14/34.

same question, the Secretary of State for Air, Lord Londonderry, stated flatly that anyone who doubted that Germany would be an enemy was dreaming, that the situation was already dangerous, and would be more so if Britain had not corrected her deficiencies by the time Germany rearmed. Again nothing was settled as it was decided that an outside committee should make a decision on the terms of reference.[47] In effect the report was shelved as the Cabinet was unwilling to come to grips with the unpleasant and expensive realities it set forth. The realities would not go away, however. There were successive reports from the Foreign Office and the subcommittee on industrial intelligence in foreign countries detailing Germany's growing capacity to produce airplanes,[48] and the disarmament conference was at the point of final collapse. On April 20 the C.O.S. submitted a paper to the Cabinet suggesting that, in view of the seriousness of the threat of German rearmament, the Opposition be called in and informed of the gravity of the situation so that they could join the Government in a campaign to win public support for rearmament.[49] This suggestion was probably more a ploy to get the Cabinet to deal with the D.R.C. report than a serious proposal. The service ministers requested that the Cabinet take a decision in principle to make armaments more efficient, and authorize the services to work out in detail the measures needed to bring them up to strength. The Cabinet decided that these questions should be turned over to the Ministerial Committee on the Disarmament Conference, D.C.M. (32), for consideration.[50]

THE DISARMAMENT COMMITTEE CONSIDERS REARMAMENT

It is more than a little ironic that the Cabinet committee on the disarmament conference should be turned over to the consideration and planning of rearmament without so much as an appropriate change in title. Its membership consisted of all the relevant and important Cabinet ministers. Ramsay MacDonald was the

[47] CAB 23/78, 10(34)1-5, 3/19/34.

[48] CAB 24/248, CP 82(34), 3/21/34, Foreign Office on Germany's illegal rearmament; CAB 4/22, 1134-B, 3/22/34, C.I.D. subcommittee on German industrial capacity for arms production; CAB 24/249, CP 116(34), 4/23/34, Vansittart on Germany's industrial capacity to produce bombers.

[49] CAB 24/249, CP 113(34), 4/20/34, C.O.S. on the German threat.

[50] CAB 23/79, 18(34)2-4, 4/30/34.

chairman, although by this time his faculties were failing, and he took little active part in the proceedings. Baldwin took MacDonald's place as chairman when he did not attend, and played a considerable role in the shaping of policy. Chamberlain of course was a dominant force in the committee. Jimmy Thomas, one of the MacDonald Labourites who joined the National Government, attended as Secretary of State for Dominion Affairs, and Samuel Hoare, the Secretary of State for India, was also a member. Rounding out the committee were the three service heads, Londonderry, Hailsham, and Eyres-Monsell. Unlike the D.R.C., this committee was made up entirely of politicians, hence was disposed to consider the political and economic ramifications of any proposed course of action very carefully. In the political climate of the times, with pacifism still a dominant force, and an election little more than a year away, this committee was not going to go out of its way to adopt programs that would upset large segments of the electorate.

At the first meeting of the D.C.M. (32) at which the D.R.C. report was considered, the question of the Far East was the first issue raised. Chamberlain took the Fisher line that it was vital to get on better terms with Japan, and that expanding the fleet did not represent the optimal use to which the nation's limited financial resources could be put. Hankey and others demurred, expressing concern for the fate of the Empire, and the discussion then shifted to the German threat. Chamberlain questioned the need for an expeditionary force to go to the Continent and defend the Low Countries. When it was pointed out that if Germany were to gain control of the Low Countries she would have an ideal staging area from which her bombers could strike every part of England, Chamberlain replied that if the Air Force were built up it would provide a deterrent to Germany's even considering such a move.[51] The C.O.S. was concerned by the suggestion that Britain could do without a force to send to the Continent, and in a memo replying to D.C.M. (32) questions, put the case in the following terms:

Question: Does a contribution towards an alliance against Germany necessarily involve an Expeditionary Force? Could it not be limited to air and naval forces?

[51] CAB 16/110, D.C.M. (32), 41st meeting, 5/3/34.

Answer: Assistance on the sea and in the air will always appear to continental peoples, threatened by land invasion, to be but indirect assistance. Refusal on our part to provide direct assistance will be interpreted by our allies as equivalent to abandoning them to their fate. . . . Unless we possess some land forces capable of early intervention on the continent of Europe potential enemies as well as potential allies will probably for the reasons given above, consider our power to influence a decision by arms inadequate.[52]

The truth of this statement was to be borne out all too clearly in 1938–39. In light of this reply and the support it received, Chamberlain shifted his ground. Claiming that he thought that the D.R.C. proposals for an expeditionary force meant creating such a force from scratch rather than upgrading a force that was already in existence, he stopped questioning the need for the force, and instead expressed the conviction that in the interests of finance the re-equipping of the force should be stretched out over a longer period of time. Moreover, he doubted that the re-equipment should begin until a more concrete picture of the requirements of the force had been developed.[53] This reluctance to allow the Army to become engaged in a re-equipment program was again based on the Treasury's concern about cost.

THE "KNOCKOUT" BLOW

Chamberlain's reference to an Air Force deterrent was not fortuitous. The Government had long been concerned about the effects of a prolonged bombing onslaught on Britain, being aware of the impact on London of the limited bombing in the First World War. On two occasions in the early summer of 1917 unopposed formations of bombers attacked London, causing considerable damage and leaving many civilian casualties. The Government drew several conclusions from this experience: the offence would have a natural advantage over the defence in the air; the ratio of casualties to pounds of bombs would be very high; and, given a more highly developed technology, it would be feasible to deliver a blow from the air that would cause panic among the civilian popu-

[52] CAB 27/510, C.O.S. reply to D.C.M. (32), 5/8/34.
[53] CAB 16/110, D.C.M. (32), 43rd and 45th meetings, 5/10 and 5/15/34.

lation, dislocate industry, and so crush morale that the nation subjected to such attack would lose all will to defend itself. This was the theory of the "knockout" blow, to which all urban areas were supposed to be vulnerable. The Air Force drew up figures, based on the London experience, and estimated that in an air attack on London there would be 1,700 killed and 3,300 wounded in the first twenty-four hours, 1,275 killed and 2,475 wounded in the next twenty-four hours, and 850 killed and 1,650 wounded during each subsequent twenty-four-hour period of bombing. The Air Force soon came to the conclusion that strategic bombing should be its central role, and by extension it assumed that every major power would arrive at the same conclusion. The Air Force experts, and the members of the Government and press who accepted their views, ignored the fact that the experience being generalized on was rather limited, and that, even with the technology that had been developed by 1934, problems of navigation, bombing accuracy, and effectiveness in a war context remained to be confronted, much less solved.[54] Only the specialists in the other services questioned the doctrine that had been enshrined as orthodoxy: the bomber would always get through. Baldwin told the public just this in a speech in 1932 to the House of Commons. He said:

> I think it well . . . for the man in the street to realise that there is no power on earth that can protect him from being bombed. Whatever people may tell him, the bomber will always get through. The only defence is offence, which means that you will have to kill women and children more quickly than the enemy if you want to save yourselves. . . .[55]

The revelation in the spring of 1934 that Germany was rapidly expanding her aircraft-production capacity touched off new concern within the Government that something had to be done to protect Britain from a "knockout" blow. Moreover, the Germans were openly flaunting their rearmament in the air, and considerable public concern was being expressed as well. Churchill, in a

[54] Charles Webster and Noble Frankland, *The Strategic Air Offensive Against Germany, 1939–1945*, Vol. I (London: H.M.S.O. 1961), pp. 35-62.

[55] *Parliamentary Debates* (House of Commons), 5th ser., Vol. 285, 11/10/32, col. 632.

speech to the House of Commons on February 7, 1934, warned that "the crash of bombs exploding in London and cataracts of masonry and fire and smoke will apprise us of any inadequacy which has been permitted in our aerial defences."[56] This speech marked the beginning of Churchill's outspoken efforts to persuade the Government to move more decisively on rearmament. In response to the growing public concern about the state of the R.A.F., Baldwin pledged in the House of Commons on March 8 that "this Government will see to it that in air strength and air power this country shall no longer be in a position inferior to any country within striking distance of our shores."[57]

This sudden clamor about the preparedness of the R.A.F. put the Air Ministry in difficulty. The proposals for the expansion that had been set forth in the D.R.C. report suddenly seemed inadequate, and were criticized as such. In the D.C.M. (32) John Simon expressed the belief that a strong Air Force was an immediate necessity and criticized as too long the ten years the report suggested to bring it up to standard. Chamberlain agreed, questioning the wisdom of allocating ten of the proposed new squadrons for overseas use while providing only an equal number for home defence. He urged that home defence be given first priority. Both Simon and Chamberlain felt that the Air Ministry was not planning adequately for the nation's defence. In reply, Londonderry stated that the program was a minimum program; he might have added that, had he proposed six months earlier the type of program now being urged, the Treasury would have been after his head. He explained that the rather long period of time suggested to complete the expansion was calculated to enable the aircraft industry to develop its capacity without burdening the R.A.F. with too many obsolete models.[58] It was a valid reply, but it did not quiet the doubts that were beginning to be expressed about the leadership at the Air Ministry.

THE TREASURY ON THE FIRST D.R.C. REPORT

The Treasury view of the proposals embodied in the D.R.C. re-

[56] *Parliamentary Debates* (House of Commons), Vol. 285, 2/7/34, col. 1197.
[57] CAB 24/267, CP 27(37), 1/22/37, Swinton résumé of parity pledges made by Baldwin.
[58] CAB 16/110, D.C.M. (32), 45th meeting, 5/15/34.

port were presented in a note to the D.C.M. (32) on June 20, under the heading of personal conclusions of the Chancellor of the Exchequer. It first pointed out that the D.R.C. report projected spending £76.8 million over the next five years, a rate of £15½ million per year. In addition the Navy was pressing for a building program costing £20½ million over the same span lifting the yearly total to £19½ million, all while Treasury revenue was shrinking. In terms of cost alone, "we are presented with proposals impossible to carry out." Then the paper turned to criticism in detail:

> Expenditure on the Army, even if no mention be made of an Expeditionary Force, bulks so largely in the total as to give rise to the most alarmist ideas of future intentions or commitments. On the other hand, the Air Force proposals for Home Defence, in which public interest is strongest, contemplate not more than the completion of a programme which was adopted as long ago as 1923.

The paper proposed "that during the ensuing five years our efforts must be chiefly concentrated upon measures designed for the defence of these Islands." To achieve this Chamberlain suggested building up the R.A.F. to the degree that it would serve as a massive deterrent to any nation's incurring Britain's wrath. The Army, on the other hand, was to have its spending limited to £19.1 million over the five-year period rather than the £40 million proposed in the D.R.C. report, because the Treasury did not believe that the Germans would be capable of waging a ground war for at least another five years. As far as the Navy was concerned, "the Committee must face the facts courageously and realise the impossibility of simultaneous preparation against war with Germany and war with Japan." Therefore it proposed that no new naval building be undertaken, saving £8 million in building costs. If these proposals were followed, the cost of the program would be £69.3 million, rather than the £97.3 million under the D.R.C. with the naval building included.[59]

The implications of the Treasury position were not lost on the members of the D.C.M. (32), who protested vigorously. Their

[59] CAB 16/111, D.C.M. (32) 120, 6/20/34, Chamberlain to the D.C.M. (32), a note on the finance of the D.R.C. proposals.

primary concern was that the proposals to limit the naval building program implied nothing less than an abdication of the Empire in the East. Eyres-Monsell said it was a course "which was not even advocated by the Communists in this country."[60] He was supported in somewhat milder terms by: Walter Runciman, the President of the Board of Trade, who was of course deeply concerned with the fate of British investments in the Empire; MacDonald, who felt that spending on the Navy was not wasteful of national resources, because the ships could be used both in the Far East or in the home waters as the need arose; Lord Hailsham, the Secretary of State for War, who pointed out that it meant putting India at risk; and Hankey and Baldwin. Hankey expressed his concerns on the subject to Baldwin in these terms:

> To speak quite frankly I am apprehensive of Warren Fisher's influence in this question. I don't think that he is a fit man or that his judgement is at its best. Moreover he has never been sound about the Navy or understood the defence question in the Pacific.

In the same letter he went on to urge Baldwin "not to allow the Cabinet to overlook the Empire aspects of naval defence and their bearing on the preservation of the Empire."[61] Vansittart felt that:

> To give up hope of the defence of our Far Eastern possessions was, to his mind, a policy of despair and defeatism, and he did not believe that the country was in so rotten a condition that they would not face up to realities.[62]

Chamberlain replied by reiterating his doubt that Britain would have to fight Japan, and stressing that the suspension of naval construction would only really be felt five years from then, as ships took that long to complete. The building program would in no way affect Britain's ability to protect the Empire in the intervening period, and that was the period about which everyone was concerned.

Lord Hailsham attacked the proposals to halve the D.R.C.'s

---

[60] Bright, "Britain's Search for Security," p. 230.

[61] Cambridge University Library, the Baldwin Papers, Bald. #1, 8/23/34, a letter from Hankey to Baldwin.

[62] CAB 16/110, D.C.M. (32), 50th meeting, 6/25/34.

recommended Army program, pointing out that, the expeditionary force aside, the spending was required just to make up the reserves of supplies dissipated in the lean years, and to equip the
existing forces with modern weapons. In short, they were
minimum requirements to meet existing deficiencies. He warned
that "at the present time, if the country went to war the Army
would not be fit to fight, and it would be mere massacre to send
them to do so."[63] Chamberlain replied that this was irrelevant if
no war was contemplated within five years and if it was accepted
that the R.A.F. was the best deterrent to war. He also mentioned
the unfortunate political effects that might result from giving the
impression that the Army was suddenly being prepared for war.
Britain's involvement on the Continent in the last war was a
nightmarish memory for the British people, and any suggestion
that the Government was even considering a repetition of it would
not be at all politically healthy. On this Chamberlain was supported
by Hoare, who was obsessed with the thought of the "knockout"
from the air and was therefore anxious that the Government concentrate on building up the R.A.F., and by Philip Cunliffe-Lister,
the Secretary of State for Colonies, who agreed that a strong Air
Force was the best possible deterrent in Europe. He was to be responsible for the molding of that deterrent within a year.

It was Stanley Baldwin who sensed and expressed the consensus at the meeting. He agreed that the naval cuts would have serious effects on Britain's position in the Far East and deplored
them. He then questioned the wisdom of committing the nation's
resources to one branch of defence, and suggested that there
would be greater wisdom as well as safety in maintaining the balance between the forces as suggested in the D.R.C. report. If it
were possible to finance such a program with a defence loan, that
should be considered; if not, cuts should be distributed evenly
among the forces. Finally he echoed the views of Maurice Hankey
when he said that both Germany and Japan "were political mad
dogs, but the scope for a mad dog was wider in the Far East than it
was in Europe,"[64] at least in the short run.

Chamberlain's closing response was that to accept the appro-

[63] CAB 16/110, D.C.M. (32), 51st meeting, 6/26/34.
[64] CAB 16/110, D.C.M. (32), 50th meeting, 6/25/34.

priations called for by the D.R.C. report would mean that the public would have to forego the tax cuts the Government had promised them, and he doubted they would react favorably at the next election even if educated on the defence situation. He then lashed out at Baldwin's suggestion of a defence loan.

> He regretted that the suggestion . . . had been put forward, as he regarded that as the broad road which led to destruction. No doubt it would be the easiest method of finding the money, since it put upon succeeding generations the onus of repaying it. He hoped that we had not yet come to that stage and would be prepared to pay our own debts in our own generation.[65]

Based on a concern for the maintenance of financial orthodoxy and a keen appreciation of the political repercussions of increased taxation, this was the Treasury position. In its view the financial and political dangers that were inherent in the increased spending suggested by the D.R.C. were a far more serious threat to the national security than the foreign menaces that the D.R.C. report sought to deal with. Their alternative was calculated to minimize the foreign threat to the home islands while maintaining the domestic economic and political status quo. The cost was the risk of the loss of the Empire—a minimal risk to their minds—and the lack of ability to intervene on the Continent, an ability for which they saw no use anyway. It was a position that time was to alter in certain particulars but that was to remain constant in its fundamental rationale throughout the thirties.

DECISION FOR DETERRENCE

The Air Force program was the first to be substantively dealt with after Chamberlain's assessment of the D.R.C. report was discussed. A subcommittee on the allocation of air forces, chaired by Baldwin, criticized the Chamberlain paper's proposal for the creation of thirty-eight new squadrons without reserves as opposed to the twenty with full reserves suggested by the D.R.C. on the grounds that this "shop window" deterrent "contemplated by the Chancellor of the Exchequer would not be capable of operating on

[65] Ibid.

a war footing for more than a week or two."[66] Moreover, the deterrent would not be effective because the enemy would realize from the output of the British aircraft industry what the situation really was. The subcommittee suggested a compromise whereby £20 million would be spent over five years to create forty and a half new squadrons, which would be largely without reserves in the first years of the program, but which would enable the aircraft industry to create new plants to supply the reserves at a later stage. The rationale behind this program was, in Baldwin's words, that it would "act as a deterrent to Germany and inspire confidence at home."[67] In that it did not exceed the financial limits for the R.A.F. suggested by the Treasury, the program was accepted by the D.C.M. (32) on July 12, 1934, and ratified by the Cabinet six days later.

The Army was not as fortunate. At a July 17 meeting the D.C.M. (32) asked Hailsham to prepare a five-year program to correct Army deficiencies within a budget of £20 million, one half the sum recommended by the D.R.C. and exactly the amount the Treasury stipulated as the limit that should be spent on the Army. Chamberlain added insult to injury by saying that the Treasury could not commit itself to guarantee even that spending level over the five-year period. This reduction had the effect of protracting the period of time needed for the Army to reach preparedness well beyond the time span for which the £20 million was allotted. The reason this decision came to pass was that Baldwin was persuaded by the Treasury arguments on the financial situation, and convinced that the R.A.F. was the most effective deterrent to Germany.

The Admiralty could do no better than to reach a stalemate in their battle with the Treasury over the expanded building program. Although its arguments about the security of the Empire had the sympathy of the Cabinet, the Treasury contentions—about the economic and political dangers involved in such a spending program, and about the building program's not having any real impact on the Far Eastern situation until it was completed

[66] CAB 27/514, D.C.M. (32) 123, 7/11/34, Baldwin subcommittee report.

[67] CAB 24/250, CP 193 (24), 7/16/34, Baldwin subcommittee recommendations to the Cabinet on the R.A.F.

five years hence—gave the Cabinet pause. The Cabinet took the politician's way out, agreeing to postpone consideration of the program until after the upcoming naval discussions with Japan and the U. S. on the continuance of the Washington and London naval agreements.

On July 31 the Cabinet approved the D.R.C. program as amended by the D.C.M. (32). In so doing it recognized and chose to deal with the burgeoning German threat in what it considered the most efficient manner, through deterrence. The Cabinet believed that if Britain showed herself willing and able to build an air force on a scale greater than any ever before undertaken, Germany would be so struck with awe that she would be discouraged from trying to compete with Britain in this arena, accepting permanent inferiority in the air. This superiority would then be the trump card with which Britain could restrain whatever outrageous excesses Germany might embark upon. Thus not only Britain's national security would be ensured, but Europe's as well. The ploy was a desperate one, undertaken more for its lack of impact on the domestic status quo than for its promised impact on the international situation. The decision to thus strengthen the Air Force marked Britain's first very tentative step toward rearmament.

ITS IMPACT ON THE BUDGET

While these deliberations were taking place, the 1934 budget was being formulated. It was a budget that was notable mostly because of the amicability with which the Treasury and the services reached agreement on the services' estimates. The R.A.F. was actually criticized for turning in too low an estimate, while the Navy settled its estimate without its characteristic high-pressure tactics. The Army felt compelled to request restoration of £1.5 million that had been dropped in the budgets of 1932 and 1933 due to the pressing requirements of day to day expenditure. The Treasury cheerfully complied, and the Army, as a token of its appreciation, returned £38,000 to the Treasury. Chamberlain, in a note on this transaction, wrote, "I think this shows the wisdom of expressing appreciation of good work."[68] This era of goodwill ended with the Cabinet's authorization of the D.R.C. program. As costs increased, so did animosity between the Treasury and the services.

The 1935 budget was affected not only by the decisions resulting from the D.R.C. report, but by new revelations in October of 1934 about German rearmament. The situation was concisely stated in a Cabinet memo of the 23rd of that month.

> From the information at the disposal of his Majesty's Government it is clear that the German Government is now in a position to come into the open at any time and confront the Powers with the fact that the German Army has been expanded to 3 cavalry and 21 infantry divisions, with a peace establishment of 300,000 men. It is also understood that the present aim of the German Government is to achieve a first-line strength of 1,300 machines [planes] by October 1st, 1936.[69]

The Government decided that the only way to deal with this blatant breach of the Versailles Treaty, short of sending troops into Germany, was to officially recognize the existence of the forces without too much moralizing in hopes of getting Germany to return to the league and agree to some type of arms limitations. At the same time the Air Ministry suggested that the Government accelerate the rate of expansion of the R.A.F. by announcing that the time for the completion of twenty-two squadrons was to be cut from four years to two. The Treasury was not at all pleased about this rather costly gesture to manifest the nation's resolve. Chamberlain asserted that there was nothing in the reports of German expansion to warrant such a move, "reminded the Cabinet of the views of the Air Ministry that their programme, as at present arranged, was as much as could be accomplished efficiently and without waste of money and effort," and pointed out that the new acceleration would cost an added £250,000 in the current year and £500,000 in the following one. He closed by stating that "the [budget] situation was a very serious one, and he felt bound to

[68] T 161/580/35164/34, 1/19/34, a handwritten Chamberlain note on the Army estimates.

[69] CAB 24/251, CP 265 (34), 11/23/34, Report of the Cabinet Committee on German Rearmament, MacDonald, chairman.

warn the Cabinet against incurring fresh commitments, having regard to the very grave difficulty in finding additional revenue to meet the increased expenditure."[70] The Cabinet ignored this counsel of caution, authorizing the acceleration at the same Cabinet meeting. It was the only time in the course of the rearmament program that the Treasury was so abruptly overridden.

Given the Government's desire to grant tax reductions and complete the return of the wages of government employees to their pre-austerity budget levels—both of which could not but be helpful in an election year—while of course balancing the budget, the increase in defence expenditure was not welcome, especially at the Treasury, which had to find the revenue to finance it. The Government did manage the increase, however, even with the increase in the R.A.F. estimate, thanks to an economy that was rapidly expanding. The only other conflict with the services arose over the Navy estimates, which the Treasury insisted be cut by £3/4 million on the grounds that the cut was needed to balance the budget. This was quickly agreed to.

THE DILEMMA

The Government's dilemma was described succinctly by Ramsay MacDonald when in the course of a C.I.D. meeting he pointed out:

> that the financial aspect of the whole question of defence preparations was extremely important; we could not run the risk of a financial smash. On the other hand if we had sufficient financial resources, it was most unwise to have insufficient defence.[71]

Torn between what it saw as its obligation to guide Britain out of the depression and back to prosperity (the task on which its political survival depended) and its obligation to provide for the national security, the Government responded by seeking to create the image of power without investing in its more costly substance. This was the rationale behind both the policy of "showing tooth"

[70] CAB 23/80, 43 (34) 2, 11/26/34.
[71] CAB 2/6, 266 (1) (a), 11/22/34.

in the Far East, and that of creating the "shop window" deterrent against Germany. It was a policy recognized as involving risks, but the risks were considered far more acceptable than those to economic recovery inherent in developing the substance of power. As deterrence was seen to fail, Britain reluctantly turned to the creation of the force necessary to defend herself. The new force, however, was to be influenced by the same restraints that shaped the old.

CHAPTER II

Towards a Substantive Commitment, April 1935–August 1936

THE WHITE PAPER ON DEFENCE

IN the White Paper on defence issued on March 4, 1935, the Government announced to the public that in view of the deterioriation of the world situation and especially the failure of the disarmament conference, the nation had undertaken to equip herself to deal with the military threats posed to the national security by foreign nations. It went on to state in general terms the needs of the armed services, and warn that the preparations envisioned would entail additional spending.[1] Little of this should have come as a surprise to anyone who had followed the defence debates in Parliament the previous year, debates in which Churchill had challenged the Government on its military preparedness, and the Government had announced successive expansions of the R.A.F. Although the White Paper elicited little in the way of response from the public at large, it touched off repercussions in Germany that caused universal concern. On March 9 Hitler used the White Paper, which he claimed, not without reason, was directed against Germany, as a pretext to announce the existence of the German Air Force to the world. A week later he reintroduced conscription, and raised the peacetime strength of the German Army to 550,000 men.

Although the Labour opposition protested the proposals set forth in the White Paper, it was not wholly united in its protest. George Lansbury, the leader of Labour in Parliament, expressed the party's official position when he declared that those proposals

---

[1] Statement Relating to Defence (Cmd. 4827, 1935). The White Paper was conceived by Warren Fisher as a means to jolt the public into the realization of the precariousness of the nation's defence posture. Written as a forceful summary of the D.R.C. report by Hankey, it was watered down by the Cabinet, which feared that it would affront Germany—which it did anyway. See Bright, "Britain's Search for Security." p. 329.

ran contrary to the tenets of pacifism, on which both his party's and his own philosophies were based. Others in the party, who felt that pacifism was no longer a viable alternative in those troubled times, had come to see collective security within the framework of the League of Nations as the only means by which aggression could be thwarted without resort to world war. Clement Attlee, who was to replace Lansbury as leader of the party within the year, stated the objections of this section of the party in the course of the March 11 debate on the White Paper.

> Nothing that I say today can be construed as suggesting that we are palliating in any way Germany's action in leaving the League, Germany's re-arming, Germany's preaching of war in Germany; but the question is, "How are you going to deal with that?" We believe that it must be dealt with not by a few nations but by the whole world. We believe in the League system in which the whole world should be ranged against an aggressor.[2]

He went on to assert that the White Paper presaged nothing less than the Government's abandonment of the concept of collective security under the League, warning that "It marks a complete change of policy. We are back in a pre-war atmosphere. We are back in a system of alliances and rivalries and an armaments race."[3] Labour's criticism failed to gain much support in the press. *The Times* in a March 15 editorial applauded the Government's resolve to strengthen the nation's defences, while deploring Lansbury's resolute pacifism as unrealistic. The same article, however, also urged the Government to continue to conduct its foreign policy within the context of collective security and the League. Few realized as Attlee did that an armaments race between Britain and Germany had been joined.

Within a month the public was shaken from its apathy when the Secretary of State for Foreign Affairs, John Simon, revealed in Parliament that, during his state visit to Germany in late March 1935, Hitler had claimed that Germany had already achieved air parity with Britain. This caused a renewal in the popular press of the

[2] *Parliamentary Debates* (House of Commons), Vol. 299, col. 40.
[3] Ibid., col. 43.

horror stories about the effects of bombing and aroused considerable public concern. That concern was increased by Churchill's resumption of his attacks on the efficacy of the Air Ministry's efforts to strengthen the Air Force, and was not wholly allayed by the Government's assurances that the nation's air defences were secure. In the course of the spring pressure built up in both the country and the Commons in support of a stronger, more efficiently managed program of rearmament in the air, pressure that was to have considerable impact on the shape of the rearmament program that eventually emerged. The Government, however, required neither the urgings of the public nor those of Parliament to galvanize it to action; Hitler's claim of parity had touched off near panic in high government circles, and drove the Government to reevaluate the whole air program.

In November 1934 British intelligence had pegged the total German air strength at between 600 and 1,000 military aircraft[4] (well below the British establishment), with a capacity to produce 140–175 aircraft per month, a capacity that was expanding all the time. After Hitler's revelation, British intelligence showed only 1,320 German aircraft as of April 1935, a figure in line with previous estimates and well below the numbers Hitler was claiming.[5] Of those 1,320 an estimated 410 were in operational squadrons, while the rest were either obsolete or reserves.[6] There were other figures about, but what is significant is that members of the Government believed, as Simon claimed to in a letter to Baldwin in April, that German rearmament in the air was proceeding faster than the Air Ministry figures showed, and that "this country is seriously open to the threat of sudden attack by a Continental Power in a degree to which it has not been exposed for hundreds of years."[7] The fact was that the Air Ministry figures on the German Air Force were accurate but their estimates of the rate at which the Germans were expanding their capacity to produce airplanes were low. At that time, however, no one was sure of the size of the German Air Force as the result of Hitler's statement,

[4] CAB 24/251, CP 265 (34), 11/23/34, paper of Committee on German Rearmament.

[5] CAB 4/23, 1173-B, 4/11/35, F.C.I. industrial intelligence.

[6] CAB 4/23, 1180-B, 6/14/35, air staff estimates of German air strength.

[7] Bald #1, 4/10/35, Simon to Baldwin, Chamberlain, and Ramsay MacDonald.

and there were a number of people in the Government with an almost obsessive fear of the "knockout" blow who were prepared to believe the worst.

AIR EXPANSION IN EARNEST

The Air Ministry responded to the situation by stating its skepticism about Hitler's figures and stressing that, even if they were accurate, Germany simply had not had the time to train and organize adequately the personnel and ground facilities to make the Air Force an effective unit. It stressed that Britain's superiority in these areas made the R.A.F. a superior force regardless of numbers. However, the Air Ministry did not hesitate to propose a new expansion program. Under this program the Home Defence Force would be expanded to 100 squadrons, an increase of 30 over the plans authorized the previous July, at a cost of an additional £17 million, with 1938 as the target completion date.[8] A hastily conceived scheme, it did not confront many of the difficulties involved. J.D.B. Fergusson, Treasury under-secretary, expressed just this view in a memo to Chamberlain in which he concluded:

> So far as I know, nobody who is acquainted with the Air Ministry has much confidence in the present direction or leadership, either civil or military. If we are really faced with a problem vital to our national safety it is lamentable that we do not have men of the highest calibre at the head of the Air Ministry and the Air Force.[9]

This view was apparently widely held, as the Cabinet appointed a subcommittee to the D.C.M. (32) on air parity to look into the situation. Its chairman was Philip Cunliffe-Lister, who within a month was to be appointed as the new Secretary of State for Air. In the first interim report the committee accepted that, to keep pace with Germany, the R.A.F. would require 3,800 aircraft by April of 1937 to give them a first line strength of 1,500 with full reserves. The projected program was now estimated to cost

[8] Ibid., 4/10/35, Londonderry to Baldwin, Simon, and MacDonald proposing new R.A.F. expansion; CAB 24/254, CP 85 (35), 4/15/35, Ellington on German air rearmament.

[9] T 172/1830, 4/27/35, Fergusson to Chamberlain on the proposed R.A.F. expansion plans.

£9,350,000 more per year than the current 1935 air estimate of £20.5 million. Significantly the Air Ministry had insisted that this new schedule of expansion could be met "without the adoption of special steps by His Majesty's Government for the reorganisation of industry."[10] This report prompted Fergusson at the Treasury to express his view to Chamberlain that the paper was a contradictory whitewash. He was especially skeptical about the suggestion that the aircraft industry could handle the expansion without reorganization, as well as about the numerical criteria for parity with Germany that the committee had chosen as its guideline. He stated, "I have the feeling, underlying both the above points that neither the Air Ministry nor the Committee have really addressed themselves to the vital issues, strategic and political."[11]

Nonetheless, on May 21, 1935 the Cabinet authorized the recommendations of the committee that the ordering of the numbers and types of planes suggested commence immediately, subject of course to Treasury review of those orders in detail.[12] The rationale behind the authorized plan, known as Scheme C, was that the purchase of 3,800 airplanes between then and April 1937 would not only enable Britain to maintain parity with Germany, but would completely occupy the productive capacity of the aircraft industry and encourage its expansion. The initial order of 3,800 planes was to be for obsolete types of aircraft, mostly bombers, because those were the only types the aircraft industry had the expertise and equipment to construct on such short notice.[13] Those planes would fulfill Baldwin's pledge to maintain numerical parity with Germany by providing what was called a shop window deterrent, while the Air Ministry and the aircraft industry undertook to plan and organize the mass production of the most modern types.

NEW MEN AT THE AIR MINISTRY

The most important step in the Government's efforts to strengthen

[10] CAB 27/518, D.C.M. (32) (AP) 3, 5/8/35, interim report on air parity, also printed as CAB 24/255, CP 100 (35).
[11] T 172/1830, 5/9/35, Fergusson to Chamberlain on D.C.M. (32) (AP) 3.
[12] CAB 23/81, 29 (35), 5/21/35.
[13] Peter Fearon in his article "The British Airframe Industry and the State—1918–1935" (*Economic History Review*, 2nd series, Vol. 25, no. 2, May 1974), provides a good account of the problems and shortcomings of the British aircraft industry up to 1935.

the nation's air defences was taken little more than two weeks after the Cabinet approved Scheme C, when Stanley Baldwin appointed Philip Cunliffe-Lister to replace Lord Londonderry as Secretary of State for Air. The appointment was part of a general Cabinet realignment that had been initiated by Ramsay MacDonald's decision to step down as Prime Minister. On June 7, 1935, bowing to the reality of his growing physical infirmity, he traded positions with Baldwin, taking the less taxing position of Lord President. Baldwin, who thus became Prime Minister for the third time, took the opportunity to move and replace those in the Cabinet who had not inspired his confidence during their tenure under MacDonald. John Simon, whose performance as Secretary of State for Foreign Affairs had attracted considerable criticism, was moved to the Home Office, and Samuel Hoare was brought in to replace him. It was, however, the appointment of Cunliffe-Lister as head of the Air Ministry that signalled Baldwin's resolve to shore up Britain's air defences.

Cunliffe-Lister, who was soon to become Viscount Swinton, had impressed Baldwin with his energy, organizational ability, and knowledge of industry when he was a member of Baldwin's subcommittee on the allocation of air forces in 1934. His appointment as chairman of the air parity subcommittee made it evident that he was Baldwin's choice to succeed Londonderry. An executive by nature, he was ideally suited to oversee the rebuilding of Britain's air force.

Joining Swinton at the Air Ministry as a special industrial adviser was Lord Weir, a Scottish industrialist and engineer who had been advising the National Government on industrial questions since 1933, and had served other Governments in a similar capacity since the First World War. A conservative who traced Britain's industrial decline to the growing power of the unions and the social welfare system that they had caused to come into being, he mirrored the views of the industrialists of his era. His primary concerns were efficiency and productivity, which he felt organized labor impeded, and he spent a good part of his career battling the unions on these issues.[14] With his contacts within industry, his knowledge of its operations, and his incisive, analytical mind he was the perfect adviser on industrial questions for a Conservative

[14] See footnote 5, Chapter I.

Government. He and Swinton were largely responsible for both the virtues and the vices involved in the expansion of Britain's aircraft production capacity, which was the focal point of British rearmament before the Second World War.[15]

Cunliffe-Lister's first task as Secretary of State for Air was to shepherd the newly authorized Scheme C through the House of Commons. He did not have an easy time, as the Labour opposition used the occasion to challenge the adequacy of the Air Ministry's plans to prevent the aircraft industry from realizing excessive profits from the sudden increase in business that the new program would send their way. Although the measure authorizing the new expansion passed without difficulty, the issues Labour raised in the course of the debate were telling ones.[16]

DEFENCE REQUIREMENTS RECONSIDERED

While the Air Force program was being debated in Parliament, the D.R.C. was reassessing its defence requirements recommendations. Moved by the deterioration of the international situation, and especially by the alarming information that Germany was borrowing as much as £1,000 million per year for rearmament, the D.R.C. issued a paper recommending a balanced expansion of the three services. While noting that "the Government have deemed it their first duty . . . to make a large increase in the R.A.F., partly as a deterrent to Germany and partly in order to secure a more rational state of public opinion," the D.R.C. expressed the belief that if real security were to be assured, all the services should be strengthened. The subcommittee also warned that "it is of the utmost importance that this country should not become involved in war within the next few years,"[17] while efforts were being made to rebuild the nation's strength. This paper was referred to the Defence Policy Requirements Committee, known as the D.P.R., which had replaced the by then badly misnamed Ministerial Dis-

[15] W. J. Reader's biography of Lord Weir, *Architect of Air Power: The Life of the First Viscount Weir of Eastwood, 1877–1959* (London: Collins, 1968) is a remarkably good work on a man whose importance has been overlooked by historians who believe that only diplomats, politicians, and intellectuals shape history.

[16] For the specifics of this debate and a discussion of the profiteering issue see Chapter III.

[17] CAB 16/138, D.P.R. 6, 6/7/35, revision of defence requirements.

armament Committee, D.C.M. (32).[18] The D.P.R. authorized the D.R.C. to re-evaluate the nation's defensive posture with reference to the increased threat posed by Germany. Chamberlain felt compelled to add that

> the financial situation does not justify the assumption that within the next few years large sums can be found for the Defence Services in addition to the total sums already provisionally approved. This need not be taken, however, to mean that all reasonable increases in expenditure are to be ruled out.[19]

On July 24 the D.R.C. issued its interim report. It stated that by all calculations it was impossible to guarantee that Germany would not become involved in a war after January 1, 1939, which was, therefore to be the target date for the completion of Britain's preparations for basic military security. In a reference to the financial restrictions Chamberlain alluded to, the committee replied that

> If . . . this country is to equip itself adequately for its own security and to discourage aggression, we can see no alternative but for the Government to widen its horizon and to resort to some system of capital expenditure for this purpose.

This was an oblique suggestion that in order to compete with Germany the Government was going to have to resort to some kind of defence loan. The committee stressed that it did not expect any kind of push for capital to be made before the election. All it asked was "authority to work out the programmes entrusted to us on the basis that they are to be completed as soon as possible—financial considerations to be of secondary importance to the earliest possible security."[20] The gist of the report was well summed up by Fisher when he said, "In our present state our country is not a deterrent to Germany; in fact we are merely a temptation."[21]

[18] CAB 16/138, D.P.R. 1, 1/7/35, terms of reference of the D.P.R. Although technically a new committee, the personnel on the D.P.R. remained the same. Its purpose was "to ensure that our defensive arrangements and our foreign policy are in line." As the ministerial committee that reviewed all the D.R.C. recommendations, the D.P.R. could report to either the C.I.D. or the Cabinet.

[19] CAB 16/136, D.P.R., 2nd meeting, 7/8/35.

[20] CAB 16/112, D.R.C. 25, 7/24/35, interim report on the defence program.

[21] CAB 16/112, D.R.C., 14th meeting, 7/19/35.

The D.P.R., meeting on July 29, considered the D.R.C. interim report and authorized the D.R.C. to proceed to draw up a program to meet the assumption of military readiness by 1939 without regard to financial limitations, to outline measures to increase industrial production of military goods, and to outline as well the state of readiness under existing programs if no special efforts were made.[22] The plan that resulted was called the Ideal Scheme.

THE IDEAL SCHEME

The Ideal Scheme was presented to the D.P.R. on November 11, 1935. In it the D.R.C. reiterated its earlier advice about the need to stay on good terms with both Japan and Germany due to Britain's current military weakness. The recommendations that were set forth envisioned: an expansion of the reserves of the Air Force in addition to the already authorized extensive plans; a two-power standard for the Navy, usually referred to as the New Standard, whereby the Navy would have a fleet sufficient to meet Japan in the Far East while still being able to deal with any threat that might arise simultaneously in the home waters; and, an Army with a Field Force along the lines suggested in the first D.R.C. deficiency report. The total projected cost of this program between 1936 and its completion in 1940 was £417.5 million above and beyond the current yearly level of defence expenditure of £124 million.[23] It was a program drawn up by dedicated civil servants to serve objectively the best interests of national security, and it was the plan the nation most certainly would have followed if economic and political considerations had been of no significance. This of course was the rub, for economic and political considerations were of the greatest significance, overshadowing even that of national security.

THE ELECTION

While the D.R.C. worked through the fall of 1935 to draw up its recommendations for the Ideal Scheme, Baldwin and Chamber-

[22] CAB 16/136, D.P.R., 4th meeting, 7/29/35.
[23] CAB 16/139, D.P.R. 52, 11/21/35, also printed as D.R.C. 37.

lain were engaged in making political plans, the success or failure of which would play a large part in determining the fate of the D.R.C.'s proposals. On October 23 Baldwin called the long anticipated election, which was to be the last, as it turned out, before the end of the European phase of the Second World War. The basis of the National Government's campaign was that it had successfully brought the country out of the depression and into an era of new prosperity, albeit a prosperity that was not shared by the 10 percent of the working population still unemployed. The economic issues were on the National Government's side, and those were the issues that would have jeopardized its chances.

To Baldwin and Chamberlain, however, rearmament was the most important overall issue in the election. Both saw the coming campaign as an opportunity to secure a solid popular mandate that would enable the Government to undertake the measures necessary to secure the nation's defences, measures that the D.R.C. was then in the process of formulating. They were split, however, over whether to make rearmament the central issue of their campaign. Chamberlain had initially suggested "that we should take the bold course of actually appealing to the country on a defence program,"[24] but Baldwin demurred. Thinking of the swings in votes against the Government in recent by-elections, and no doubt recalling the disaster that befell the Conservative candidate who had advocated the strenghthening of the armed forces in the October 1933 by-election in East Fulham, he felt that the risk of confronting the still potent pacifist sentiment in the country was too great to take.[25] Moreover, he realized that even if the National Government were to win overwhelmingly at the polls, if rearma-

[24] Feiling, *Chamberlain*, p. 266.

[25] The conclusions Baldwin, the Conservative Party, and the press chose to draw from the results of the East Fulham by-election in fact bore little relation to what actually transpired in the contest itself, as C. T. Stannage and R. Heller both point out in their respective articles "The East Fulham By-Election, 25 October, 1935," *Historical Journal*, 14, no. 1 (1971), pp. 165–200 and "East Fulham Revisited," *Journal of Contemporary History*, 6, no. 3 (1971), pp. 172–196. Martin Ceadel puts the election in its broader perspective in his essay "Interpreting East Fulham," which appears in *By-Elections in British Elections* (New York: St. Martin's, 1973), edited by Chris Cook and John Ramsden. East Fulham is a classic example of what was perceived to have happened being more important than what actually took place, of myth creating its own reality.

ment had become the devisive issue in the campaign, any chance there might be of uniting the nation behind an expanded defence program in the immediate future would be jeopardized.[26] Therefore it was decided that the Conservative Party would stress the importance of correcting the deficiencies in the nation's defences, but minimize the scale of the anticipated corrective programs and the cost they would entail. The party manifesto, drafted by Neville Chamberlain, set the tone and context in which the party's candidates discussed rearmament.

> The Covenant itself required that national armaments should be measured both by the needs of national defence and by the duty of fulfilling international obligations . . . our influence can be fully exerted only if we are recognised to be strong enough to fulfil any obligations which jointly with others we may undertake. . . . We have made it clear that we must in the course of the next few years do what is necessary to repair the gaps in our defences which have accumulated over the past decade. . . . The defence programme would be strictly confined to what is required to make the Empire and the country safe and to fulfil our obligations towards the League.[27]

By emphasizing the need for rearmament while playing down the scope of the programs that they even then knew would be required, the Government hoped to alert the electorate to its importance without causing alarm that would hurt it at the polls.

Labour opposed the minimal rearmament program that the Government set forth, although not in terms of resolute pacifism as it had done up to this time. At its annual meeting at Brighton shortly before the election was called, the Labour Party had declared that it would:

> efficiently maintain such defence forces as are necessary and consistent with our membership in the League; the best defence is not huge competitive armaments, but the organisation of collective security against any aggressor and the agreed reduction of armanents everywhere.[28]

[26] Middlemas and Barnes, *Baldwin*, p. 867.
[27] Ibid., p. 866.
[28] *Parliamentary Debates* (House of Commons), Vol. 307, 12/5/35, col. 330–331.

Passed only after a bitter internecine struggle, this declaration represented a retreat from Labour's traditional pacifist position, a retreat that had been presaged by the remarks of Attlee and others in the course of the debate on the Government's White Paper on defence six months earlier. As a consequence of this turn from the doctrine in which he so strongly believed and for which he had for so long and determinedly fought, George Lansbury surrendered his leadership of the party to Attlee.

Although Labour's new position was identical on the face of it to that taken by the Conservatives, Labour continued to oppose the Conservatives' rearmament proposals on the grounds that they exceeded the level of armaments that would be required if defence planning were really done within the context of collective security and the League of Nations. In the course of the campaign, Labour candidates charged that their opponents' defence program was in fact only the first step down the road to massive rearmament. Deriding the Conservatives' professed intention to conduct rearmament within the context of collective security and the League as a sham, they warned that if the National Government was returned to office the nation would soon find itself in the midst of a competitive arms race like that which preceded the First World War.

While Labour's warnings were more prescient than it knew, the logic of its opposition to any new armaments was rendered less than convincing by its position on the unfolding Ethiopian crisis. After Italy's October 3, 1935, attack on Ethiopia, the question arose whether the League should impose sanctions against her. While both Labour and the Government took positions favoring sanctions, the Government used the issue to point out the inconsistency of Labour's supporting sanctions yet opposing measures that would give the nation the military strength to enforce them. It was a telling point, although it was more than a little disingenuous. The Government's resolute public position on sanctions was taken largely for political reasons as the Hoare-Laval affair soon revealed.

That, however, was in the future. When the nation went to the polls on November 14, it returned the National Government with a commanding majority only somewhat smaller than that by which they had won in 1931. The Conservatives won 387 seats, and were joined by 22 Liberal Nationals and 8 National Labourites. Labour, in a respectable recovery from the disaster of 1931, won 154 seats. This result was the National Government's reward for making good its promise to deliver the nation from the depression.[29]

BALDWIN AND THE HOARE-LAVAL AFFAIR

Fortune did not permit Baldwin to savor for long his Government's victory at the polls. His hopes that he would have a unified majority behind him as he undertook to bring the nation's defences up to the standard required to meet the increasingly ominous German and Japanese threats were soon dashed by the Hoare-Laval affair. His handling of that crisis damaged both the country's and his party's confidence in his leadership. The Ministry of Supply debate that followed on the heels of the affair exposed the divisions within his party and the erosion of his political control. Baldwin's failure to provide decisive leadership on these occasions had two important effects: first, by undermining his control over his Cabinet it diminished his ability to determine the shape and direction of the new rearmament program; second, by exposing the divisions within the Government over the formulation of policy it impaired efforts to rally the nation behind the defence program. There can be no question but the failure of Baldwin's political grasp contributed to the lack of coherence and organization that plagued the first year of British rearmament.

The Hoare-Laval Affair, which initiated Baldwin's misfortunes, had its roots in the Government's handling of the Ethiopian crisis. At the heart of that crisis was Italy's longstanding desire to establish a colony in East Africa at Ethiopia's expense. By early 1935 the Italian dictator, Benito Mussolini, had determined that the time was right to seize the territory that Italy had coveted for so long.[30] The British Foreign Office struggled through the spring and summer to persuade Mussolini not to pursue his goals through

[29] Mowat, *Britain*, pp. 553–556.

[30] The following summary of the Ethiopian crisis and the Hoare-Laval Affair is drawn from the detailed account in Middlemas and Barnes, *Baldwin*, chapters 30–31. See also R.A.C. Parker, "Great Britain, France and the Ethiopian Crisis, 1935–36," *English Historical Review*, no. 89 (April 1974), pp. 293–332; and in the *Journal of Contemporary History*, Aaron Goldman, "Sir Robert Vansittart's

the use of military force, but to no avail. Britain's desire to fore-stall what would have been naked aggression against a member of the League of Nations was based on the realization that such a belligerent act would force the League, and Britain as the leading military power in the League, to take action against Italy, a situation the Government wished to avoid. At the heart of the Government's reluctance to antagonize Italy was the knowledge that such antagonism would complicate and exacerbate Britain's already difficult defence problems. The Admiralty, the service that would necessarily bear the brunt of such a conflict, warned that although there was no question that Britain would prevail in a confrontation, the cost in terms of ships lost would be completely unacceptable in view of Britain's already overextended global naval commitments. The admirals went on to point out that even if the situation did not degenerate into outright conflict, a hostile Italy was in an ideal position to threaten and if necessary cut off Britain's vital route through Suez to the Far East, thus further weakening her already tenuous strategic position there.

At the Foreign Office, Vansittart, the Permanent Under-Secretary, was concerned that, in the event of a confrontation with Italy, Germany and Japan would take advantage of the situation to embark on expansionist ventures. Moreover, he pointed out that there was the danger that Mussolini might feel driven to seek some sort of rapprochement with Hitler in the face of British hostility, even if the situation did not reach the point of conflict. In that event Britain would be faced with the threat of simultaneous war with three great powers, a situation, the services warned, that should be avoided at all costs. As the crisis developed, the Foreign Secretary, Samuel Hoare, came increasingly to share Vansittart's views. As a consequence of these problems, the Admiralty and the Foreign Office, as well as the Board of Trade, were unwilling to permit the Government to take any steps that would place Italy in the role of an adversary. They contended that Britain's obligations as a member of the League should not be permitted to jeopardize her national security.

Search for Italian Cooperation Against Hitler, 1933–1936," Vol. 9, no. 3 (July 1974), pp. 93–130; and James C. Robertson, "The Hoare-Laval Plan," Vol. 10, no. 3 (July 1975), pp. 433–464.

There were, however, others in the Cabinet who were not so willing to abandon the League as an instrument for world peace. Among them were Anthony Eden, at that time Minister without portfolio for League Affairs, Neville Chamberlain, and Baldwin. None questioned the strategic implications of alienating Italy, but each believed that the situation should be worked out through the League, preferably without resort to coercion. Eden argued that success on the part of the League in settling the Ethiopian issue would strengthen its credibility in dealing with similar situations in the future, thus serve as a deterrent to any expansionist intentions Germany might have. Baldwin felt that a solution should be attempted through the League because, in his biographers' words, "If it succeeded, it would be a most potent instrument against German ambition; if at length it failed, that very failure would help to educate public opinion and free the Government from future constraint."[31] From the outset of the crisis he had pledged Britain's support for the League, support that was continually reiterated in the course of the election campaign. He always added, however, that he would not involve Britain in any unilateral use of force, meaning that Britain would not act, even on behalf of the League, unless other nations acted with her.

The only other power in the League that could effectively contribute to an effort against Italy was France. Chamberlain, who fully shared Baldwin's views, recognized clearly that France held the key to a peaceful Ethiopian settlement. As early as July 5, 1935, he wrote in his diary:

> If we and France together determined that we would take any measures necessary to stop him [Mussolini], we could do so, and quite easily. . . . If the French would agree to play their part, the best way would be to go privately to Mussolini and warn him of our views and intentions, at the same time assuring him of our desire to save his face and get him some compensation from the Abyssinians. If the French would not play, we have no individual (as opposed to collective) obligations, and we should not attempt to take on our shoulders the whole burden of keeping the peace. But if in the end the League were

[31] Middlemas and Barnes, *Baldwin*, p. 877.

demonstrated to be incapable of effective intervention to stop this war, it would be practically impossible to maintain the fiction that its existence was justified at all.[32]

In the event, Chamberlain's observations presaged the policy the Government resolved to pursue, and accurately predicted the difficulties that it encountered.

The armed services had informed the Cabinet that if it should become necessary to engage Italy militarily, the support of the French, especially the availability of their Mediterranean ports, was of vital strategic importance. They stressed that without French assistance the fleet would have no safe harbors available to it for repairs in the area of combat, and this, they warned, would increase the number of British losses that could be expected. In their view French support was absolutely necessary in any encounter with Italy.

When the French were approached on the matter, they proved entirely unhelpful. What the British desired, as Chamberlain indicated, was French agreement to support an ultimatum to Mussolini that would compel him to settle the Ethiopian issue peaceably or face the combined military wrath of Britain and France. The French government, led at the time by Pierre Laval, was confronted with severe domestic economic and social instability, and had no desire to do anything that might exacerbate the situation. The British invitation to run the risk of becoming involved in a war with their Italian neighbors for the sake of a semi-feudal African kingdom, or even for the League of Nations, was not one they could be expected to welcome. Laval, however, did not rebuff the British proposals outright, but chose rather to suggest that, if a formula could be found that was certain to be acceptable to Mussolini, the French would support it.

This French reluctance to involve themselves in the problem placed the Government in an awkward position. Although Baldwin, by his pledge that the Government would not become involved in any unilateral use of force, left an opening through which Britain could escape shouldering "the whole burden of keeping the peace" in the event of French non-cooperation, it was

[32] Feiling, *Chamberlain*, p. 265.

not politically feasible to even consider such a course with the election campaign in progress. Additionally, both Baldwin and Chamberlain appeared to be genuinely anxious to effect a solution through the League, being unwilling to let it lapse into impotence without making an attempt to prove it effective.

The Foreign Office, therefore, developed what Middlemas and Barnes refer to as a double policy that they hoped would see the Government through the crisis without jeopardizing either the Government politically or the nation strategically. The public aspect of this policy involved the Government's continued outspoken opposition to Italy's belligerent action and full support of League attempts to curb it. As a consequence of Mussolini's October 3 invasion of Ethiopia, the Government publicly proclaimed its willingness to agree to sanctions against Italy, and in Geneva Anthony Eden was one of the driving forces behind the provisions for limited sanctions adopted by the League in late October. On the private side the Foreign Office continued to try to produce a proposal for the peaceful settlement of the Ethiopian question that would be acceptable to Mussolini and, as a consequence, to the French, yet would not be rejected by the other members of the League as a reward to Italy for her aggression against a League member; it also continued to try to persuade the elusive Laval to take a more resolute position.

The successful outcome of the election did little to ease the Government's predicament. The question of the imposition of oil sanctions had been raised in the League, and there was widespread support among members for them. The British, however, were discomfited by the proposal because they knew that they were the only power in the League with a navy large enough to enforce such sanctions. It was the situation the Admiralty had warned against, but there was really no viable way that the British could refuse to take part in such an action after having taken the lead in bringing pressure to bear on Italy up to this point. With the adoption of oil sanctions imminent, the Government desperately redoubled its efforts to find a solution to the issue that would spare it the risk of military involvement with Italy. The only hope was Laval and the French. At Geneva the British succeeded in delaying the date on which oil sanctions were to be imposed, while in

Paris British diplomats made every effort to win some kind of French backing for a negotiated settlement. Meanwhile, in London the resolve of the leading members of the Cabinet to see the issue through began to waver.

On December 7, en route to a vacation in Switzerland, a weak and tired Samuel Hoare stopped in Paris accompanied by Vansittart. Their intention was to have discussions with Laval that they hoped would lead to some kind of agreement. In their first meeting Laval firmly stated that France would not be a party to any conflict Britain might become involved in with Italy as the result of the imposition of sanctions. He agreed, however, to try to persuade Mussolini to settle the issue peacefully. The following day Laval presented Hoare with a formula for the settlement of the Italo-Ethiopian dispute that he believed would be acceptable to the Italians. In fact, the formula had been worked out by the Italians and Laval secretly during the night, and ceded to Mussolini all he had set out to acquire. Hoare consulted with Vansittart, whose strong views on the need to reach an amicable agreement with Rome were well known, and decided to initial the proposals, accepting their terms on the condition that they were to be ratified by the Cabinet and the League. The communiqué announcing the agreement was worded in a way that implied that the Cabinet's ratification of the proposals would be a formality. Such, however, was anything but the case.

Although Hoare had initialed the agreement on Sunday, it was not until Monday morning that London received an outline of its content. It soon became evident that Hoare had conceded more than the Government's leaders had intended to give up, and that the terms would clearly be unacceptable either to Ethiopia or the League. When the Cabinet met at 6:00 p.m., it decided that the terms of the agreement should be presented to Ethiopia and Italy simultaneously, and that they should be set before the League Council, where it was hoped they could be amended into a more acceptable form. In the course of the meeting there was a great deal of criticism of the terms, and the question was raised as to whether it would not be best to repudiate the whole thing outright. That, however, would have been tantamount to a condemnation of the Foreign Secretary, who was not present to defend himself, and might have precipitated a Cabinet crisis. As Baldwin observed, "If we disowned Sam, the French would be angry and would say we had let them down, so we backed him. We did not like it at all but the alternative seemed to us to be worse."[33] The Government resolved to make the best of the unhappy situation.

Its task was made no easier by the press, which, having obtained the nominally secret terms, published them the following morning along with scathing denunciations. That afternoon Baldwin went before a hostile House of Commons to explain the Government's actions. There he tried to smooth over and evade the issue by implying that he was not then at liberty to tell the whole story, making the unfortunate statement at one point that his lips were "not yet unsealed," and assuring the House that if he were free to make his case he could "guarantee that not a Member would go into the Lobby against us."[34] It was a singularly feeble performance that only accentuated the untenability of the Government's position. If anything, it served to swell the mounting torrent of criticism that was pouring in from every quarter. The public, the press, and even the Conservative backbenchers were outraged at the Government's apparent willingness to reward Mussolini's aggression. The Conservative's chief whip in Parliament told Baldwin outright, "Our men won't stand for it." As the week wore on, opposition to the Hoare-Laval agreement grew stronger and stronger.

When the Cabinet met again on December 18 to consider the situation, most of the members had concluded that the only feasible way out of their predicament would be to repudiate the agreement and ask Hoare for his resignation, thereby settling the blame for the affair on him. Baldwin was reluctant to turn on a close friend and trusted colleague in this manner, but in the face of the threat of half the Cabinet to resign he relented. Hoare tendered his resignation, and on the following day both he and Baldwin made their explanations in speeches to the House. Hoare's was lucid and dignified, but Baldwin's was little more than a sullen, ineffec-

---

[33] Middlemas and Barnes, *Baldwin*, p. 886.

[34] Ibid., p. 887. It is suggested by Middlemas and Barnes that the secret Baldwin could not reveal was the Secret Service's belief that Mussolini had bribed Laval to press his terms on Hoare. To have thus impugned the integrity of the French Prime Minister at that point would have hardly improved the situation.

tual apology. Austen Chamberlain, the most prestigious of the Conservative Party's senior statesmen, remarked in private, "Had I thought it compatible with the public interest, I believe that after S. B.'s miserably inadequate speech and the initial blunder, I could have so reduced his majority as to force his resignation."[35] Despite Baldwin's replacement of Hoare with Eden, who by his actions at Geneva had emerged from the affair with his reputation enhanced in the public eye, confidence in the Government, and especially in the Prime Minister, was greatly diminished. Moreover, as Austen Chamberlain's statement indicated, Baldwin's standing in his own party had received a severe blow as a consequence of his handling of the affair. His reduced stature would have a noticeable effect on his ability to coordinate the implementation and conduct of the rearmament program during the following year.

THE MINISTRY OF DEFENCE DEBATE

Criticism of the Government's conduct of defence policy followed close in the wake of the Hoare-Laval affair. Early in January articles questioning the effectiveness of the existing organization for defence planning appeared in several of London's leading newspapers. The most penetrating of these were written by Basil Liddell Hart, *The Times'* military correspondent, and they touched off a flurry of controversy.[36] Demands were made in the press, as they had been on several occasions since the war, for the creation of a ministry of defence, and these were soon taken up by Conservative members of Parliament. The critics argued that, under the existing structure, with three independent services each responsible for its own planning and supply, there was bound to be inefficient duplication of effort in certain areas, and often uneconomic competition for common stores. A ministry of defence the critics contended, would diminish the overlapping of function and effort, and allow economies to be realized through the joint purchase of common stores. On a more important level, the critics felt that such a ministry would foster the development of a more

[35] Middlemas and Barnes, *Baldwin*, p. 896.
[36] Stephen Roskill, *Hankey: Man of Secrets*, Vol. III (London: Collins, 1974), p. 202.

unified and well-integrated defence policy by forcing the commands of the three services to work together rather than compete, as they did under the existing system.

The emergence of this criticism was not fortuitous. It had been initiated by a small group of men, most of whom had some connection with the Air Force, and all of whom were convinced that the existing organization of defence planning gave the Admiralty too much influence over the formulation of policy. Led in Parliament by two Conservative former officers, Wing-Commander Archibald James and Colonel A. J. Muirhead, this group also had the active assistance of the former Chief of the Air Staff, Lord Trenchard, as well as the clandestine support of the current Permanent Secretary of the Air Ministry, Christopher Bullock. It was Lord Trenchard who initiated the press campaign with a letter in the December 16, 1935, edition of *The Times* that was highly critical of the existing defence organization and of its creator, Maurice Hankey. Through influential friends in the press like Robert Barrington-Ward, the assistant editor of *The Times*, the Air Force group had been able to bring the issue of defence organization, and especially the question of the desirability of a ministry of defence, to the public's attention.[37]

The Air Force group raised the subject at this particular time because it realized that in the next few months decisions determining the roles of each of the services in the new national defence scheme would be made, decisions that would dictate the percentage of the defence budget allocated to each service in future years, and, consequently, the scope of the programs each would be permitted to undertake. Concerned that the Air Ministry lacked the necessary influence within the prevailing policymaking structure to insure that the Air Force would be assigned a sufficiently important role in the new defence scheme, the Air Force group initiated the press campaign in order to precipitate some changes in that structure that might work to the Air Ministry's advantage.

Although the creation of a ministry of defence became the focal point of the public debate that ensued, the Air Force group had a more limited objective in mind. It wished to diminish or eliminate

[37] Roskill, *Hankey*, pp. 208–209. Roskill provides a good account of the ministry of defence debate from Hankey's point of view on pages 202–213.

Maurice Hankey's power in the existing defence organization. A former naval officer, his position as the Permanent Secretary to both the Cabinet and the C.I.D. gave him enormous influence in the formulation of defence policy, influence the Air Force group felt he used consistently to the detriment of the Air Ministry's interests. The group's purpose in initiating the press campaign was to arouse sufficient public and parliamentary concern about the organization of defence to compel the Government to make some changes in it to ameliorate that concern. It hoped that those changes would prompt Hankey, who was known to be extremely jealous of his prerogatives, to resign, or would at least dilute his influence. The Air Force group felt that any diminution of Hankey's power would result in a proportional increase in the Air Ministry's.

As we have seen, the question of the higher organization of defence moved very quickly from the press to the floor of Parliament, where it was taken up and debated by members on both sides of the aisle with great seriousness. Few were aware of the ulterior motives of those who had brought the issue to their attention. Most members knew only that there was cause for concern about the state of the nation's defences, that Britain might be vulnerable. The doubts many members had about Baldwin's fitness to provide the nation with adequate leadership—doubts that had been aroused by the Hoare-Laval affair—heightened their concern about the effectiveness of his Government's efforts to shore up the nation's defences. Even Conservative backbenchers shared these doubts. The dissatisfaction they expressed about their Government's defence organization was a manifestation of their lack of confidence in Baldwin's leadership. Had it not been for the Hoare-Laval affair, it is doubtful that the Air Force group's campaign would have amounted to anything more than a few newspaper articles.

As it was, however, a private member's bill proposing the creation of a ministry of defence was given a full day's debate in the House of Commons on February 14, 1936. Although little of import was said on the substance of the issue, as the Government made a sound case against such a ministry, the extent of the Conservative dissatisfaction with Baldwin's leadership was made distinctly evident. In a speech of devastating directness Austen Chamberlain, the party's elder statesman, expressed the feelings of many of the Government's supporters in the following terms:

> I want to direct attention to the question of what underlies the anxiety—which I think has been so obvious in every speech today—as to whether we are setting about these defence problems in the right manner. I am not surprised at this anxiety. The House and the country have experienced in recent times some rude shocks for which they were wholly unprepared, which remain to this day in large part an unexplained mystery and which make us anxious lest they should be repeated in more critical circumstances when they might be almost fatal. I am going to recall to the Prime Minister, for the purpose of illustrating what I mean, one or two of his own utterances.

Recalling Baldwin's assurances to the nation in November 1934 that Germany was not overtaking Britain in air strength, and his subsequent admission of error the following May, he asserted:

> I recall no comparable pronouncement by the head of a Government on a fundamental issue of defence in the forty odd years of my Parliamentary experience. Is it to be wondered at that some of us who are not alarmists, some of us who had a large measure of responsibility up to a certain point for not advancing our own preparations, now feel profoundly anxious? Yes, but that is not all. There is a later speech by my Rt. Honourable Friend. In the course of our debate last December, my Rt. Honourable Friend startled the House and the country by use of language such as none of us had heard in our experience from a Minister of the Crown. He said, on 10th December, 1935: "I have seldom spoken with greater regret, for my lips are not yet unsealed."—He was defending the policy of the Hoare-Laval proposals. . . . Well, Sir, why do I cite these things? They are not pleasant; they are not reassuring. It is because those things could not have happened if the thinking machine of the Government was working properly, if their defence organisation was really efficient.

He then proceeded to dismiss the idea of a ministry of defence as a solution to the problem, suggesting instead certain limited changes in the defence organization, before closing by noting: "The speeches which I quoted are our justification for demanding great changes and evidence that everything is being done to prevent a continuance or a recurrence of such events as those for which the Prime Minister had twice stood at the Table of this House within the last two years, to ask the pardon of the House."[38] The import of Chamberlain's words was not lost on either Baldwin or the Cabinet.

Although the Government was probably not aware before the February 14 debate and Austen Chamberlain's speech of the depth of the disquiet within the ranks of its followers concerning the Government's leadership, it had recognized soon after the press debate over the organization of defence erupted that it would have to be prepared to deal with demands for a ministry or at least a minister of defence. Maurice Hankey was the first to respond to the attacks on his organization. He explained that a ministry of defence, far from being more efficient and economical, as the critics alleged, would simply impose another layer of bureaucracy on those already extant, thereby diminishing the defence establishment's efficiency. He countered the arguments for the creation of a minister of defence by pointing out that such a minister, by virtue of his assumption of all the Prime Minister's defence-related duties, would be in a position to usurp the Prime Minister's control over defence policy and challenge his leadership in the Cabinet on defence matters. He added his doubts that any one man had the capacity to successfully administer all three of the services. Hankey, whose vested interest in the maintenance of the existing system has already been noted, was supported by Warren Fisher, who felt that a ministry of defence would be a threat to Treasury control. The Admiralty also opposed change because, as the most politically effective of the services, it felt the Navy could garner more funds under the existing structure than it would be able to under a ministry of defence.

Not all the members of the Cabinet agreed that the defence organization was such a perfect piece of work, although all did concur with Hankey's basic contention that the creation of a ministry,

---

[38] *Parliamentary Debates* (House of Commons), Vol. 308, 2/14/35, col. 1360–1362.

or even a minister of defence, would cause more problems than it would solve. The Cabinet recognized that the existing arrangements did have certain faults, the worst of which was that the Prime Minister, who was the chairman of the C.I.D. and the C.O.S. as well as of other defence-related committees, was becoming unable to keep up with the increasing volume of work. A second problem, raised by both the Secretary of State for Air and Chamberlain, was that of the inability of the C.O.S. to make hard decisions calling for the ordering of priorities among themselves. Swinton asked rhetorically, "Is compromise to avoid difficulties rather than face them unknown?" The Cabinet warmly acknowledged Hankey's importance to the whole defence-planning process, but noted that "there is, however, no second Hankey in the world."[39]

These issues were all raised in memoranda, cabinet papers, and letters from Cabinet members to the Prime Minister, most of which were submitted to enable the Government to prepare its defence in the February 14 debate. Until that debate, the Cabinet had not seriously considered changes in the defence organization; however, within hours of its conclusion, Hankey wrote his ally Warren Fisher that he was "afraid we have got to make some concession for a Minister of Defence," and sent Baldwin a note to the same effect along with an outline of changes in the defence structure that would silence the parliamentary critics.[40] The Ministerial Committee on the Co-ordination of Defence was quickly formed to deal with the issue, and on February 20 it presented its recommendations to the Cabinet. Based primarily on Hankey's proposals, those recommendations focused on the appointment of a minister to oversee all defence questions. His primary role would be to serve as a coordinator of defence policy, sitting on all committees dealing with defence matters and acting as a surrogate for the Prime Minister when he could not attend. One of his important tasks would be to keep an eye on the work of the C.O.S., insuring that it faced up to hard issues rather than passing them on.[41] In

[39] CAB 24/260, CP 37 (36), 2/10/36, Swinton on a ministry of defence.

[40] Roskill, *Hankey*, p. 204, refers to CAB 21/424.

[41] CAB 24/260, CP 51 (36), 2/20/36, memorandum of the Ministerial Committee on the Co-ordination of Defence.

effect this minister was to be a trouble shooter in the defence organization, making it more efficient without altering its basic method of operation or upsetting the prevailing balance of power within it. That the Cabinet approved the committee's recommendations without demur on February 24, 1936, is therefore not surprising.[42] The rapidity with which the members arrived at this solution, and its noncontroversial nature, both indicate that the Cabinet was more interested in mollifying parliamentary and public opinion than in substantially altering the defence organization. The Cabinet was well aware that any attempt at the latter would entail bitter interdepartmental struggles that would in all likelihood prove counterproductive.

It was only after two speculation-filled weeks that Sir Thomas Inskip was appointed to fill the position that had been given the rather ponderous title of Minister for the Co-ordination of Defence. This appointment was not received with acclaim in Parliament, where it had been hoped that a man of known ability and prestige would be named. Inskip had been the Attorney-General but had little administrative experience. The witticism that made the rounds in the lobbies was that there had been no appointment like it since Caligula had made his horse consul.[43] The hope had been that Churchill would be asked to take the position, but that would have created political problems in a government not committed to rearmament on the scale that Churchill felt was necessary. The only man of prestige who could have taken the post without creating difficulties was Neville Chamberlain, who did not want it. Although Inskip has been criticized by both contemporaries and many historians (who are sometimes prone to parrot the opinions of contemporaries), a thorough reading of the documents on defence covering the period during which he was Minister for the Co-ordination of Defence reveals that he did an effective job, especially considering the handicaps of his position.

THE IDEAL SCHEME CONSIDERED: THE TREASURY VIEW

In these months following the Government's successful showing

[42] CAB 23/83, 9 (36) 1, 2/24/36.
[43] Mowat, *Britain*, p. 570.

at the polls, during which the public's confidence in its leadership was profoundly shaken by the Hoare-Laval affair and its aftermath, the Government began in earnest to formulate its plans for strengthening the nation's defences. In mid-November 1935 the D.P.R. took up consideration of the recommendations made by the D.R.C. in the Ideal Scheme, recommendations which, it should be recalled, were made without reference to financial limitations. Predictably the Treasury had much to say about the expense entailed by those proposals.

At the heart of the matter was the question of how this new expenditure was to be financed. The D.R.C.'s interim report had drawn criticism from the Treasury by suggesting that a defence loan would be the best way to handle the problem, and subsequently the committee added its opinion "that we could not compete on a taxation basis against nations who were acting on a loan basis."[44] The Treasury continued to oppose the suggestion despite Warren Fisher's advocacy of it. In fact this difference between Fisher and his colleagues marked the beginning of the decline of his influence over policy.[45] Chamberlain, who also strongly opposed the idea of a defence loan, came to rely increasingly on the advice of Edward Bridges, a Treasury undersecretary, and Richard Nind Hopkins, the Second Secretary. Both were primarily concerned with economic matters, and were strictly orthodox in their financial views. Hopkins was in close touch with Montagu Norman at the Bank of England, and was very aware of the prevailing concerns of the City. The Treasury's top economic analyst was Frederick Phillips, another undersecretary, who provided Hopkins and Bridges with the information on which they based their recommendations to the Chancellor. His analysis of the implications of programs and policies provides the most penetrating insights into the rationale behind Treasury policy available anywhere in the Treasury files.

[44] CAB 16/112, D.R.C., 13th meeting, 7/11/35.

[45] By 1938 Fisher's influence had waned entirely. In May of that year Thomas Inskip, the Minister for the Co-ordination of Defence, related to Hankey that Chamberlain had told him that "knowing how Warren Fisher got worked up about things, he [Chamberlain] had instinctively a complete mistrust of any suggestion he made—so much so that, even if Horace Wilson, whose judgement he normally trusted, shared Fisher's view, he would mistrust it." Churchill College Library, HNKY 1/8, Hankey's Diary, 5/16/38, p. 96.

ON BORROWING FOR DEFENCE

In the initial response to the suggestions that a defence loan was necessary, Richard Hopkins wrote Fisher and Chamberlain that "the only ground for borrowing . . . is that the expenditure places the Exchequer seriously in deficit *when the country is taxed to full capacity.*" He went on to say that "it would be unfortunate if the country began to think of a Defence Loan as a comfortable Lloyd Georgian device for securing not only larger forces, but also lower estimates, budget surpluses and diminishing taxation."[46] After the Ideal Scheme was articulated, the Treasury began to consider seriously how such an expensive program could be financed without doing damage to the economy. Phillips' memorandum on the subject laid the foundation for the Treasury position. He pointed out:

> The economic disadvantage of heavy expenditure on armaments is that it reduces the national wealth. Goods and services are used up and the loss is permanent since expenditure on armaments will do nothing to increase the supply of goods and services in later years. . . . Heavy armaments expenditure is particularly dangerous to the capitalist states of Western Europe with their depressed incomes, their high taxation and their excessive national debts.

This is the basic classical economic view of armaments expenditure. Phillips continued:

> If we had the free choice to decide, not only for ourselves but for others, in what way any unavoidable increase in armaments expenditure was to be financed, we should undoubtedly decide that it should be financed out of revenue. For, with the exception of the U. S., there is no country in the world that would have less difficulty than ourselves in providing for armaments by that method.[47]

Unhappily Britain had no such choice. Phillips then stressed that if the country were to resort to borrowing, "It must be supposed to be clear beyond doubt that the level of expenditure con-

[46] T 161/688/14996/1, 10/7/35, Hopkins to Chamberlain and Fisher on defence finance.
[47] T 171/324, 11/29/35, Phillips on defence finance..

templated is abnormal and non-recurrent and beyond the ability of the country to discharge."[48] In concrete terms he recommended that the yearly cost of maintaining the defence services, exclusive of the non-recurrent expenditure for new guns, planes, tanks, etc., should be projected, and that amount should be budgeted to be raised out of revenue, as this constituted the level of recurrent expense, which under the precepts of orthodox finance always had to be paid for from revenue.

In a note dated December 2, 1935, based on Phillips' memorandum, Richard Hopkins attempted to predict the financial effects of the Ideal Scheme, if it were adopted, on the assumption that the nation would endeavor to continue to follow the principles of sound finance. He projected the yearly cost of maintaining the defence services after the Ideal Scheme had been completed, maintenance including debt service and obsolescence, and arrived at the figure of £215 million, an even £90 million more than the current military budget. This would be the level of defence spending to be met out of revenue in 1941 if the budget for that year were not to be in default. Hopkins emphasized that the magnitude of the sums themselves did not justify borrowing, "any more than the magnitude of the sums required at a previous time for unemployment relief then justified that course."[49] He conceded, however, that the gigantic sums for non-recurrent expenses could not be raised out of revenue "without an unbearable burden upon the taxpayer," and went on to indicate that

> even if the strictest canons of finance suggest that that burden ought to be shouldered there are other courses which, if frankly admitted to be deviations from our severest standards, would certainly be accepted by reasonable opinion at home and ought certainly in present conditions to be condoned by reasonable opinion abroad.

[48] T 161/688/14996/1, 11/29/35, Phillips on defence borrowing.

[49] Hopkins in a later note, T 171/324, 4/3/36, explains why the City was in fact more amenable to borrowing for defence than it was to borrowing to pay for poor relief in 1931: "The real answer lies in a rather different (less quotable) plane. In those eyes which mattered so much when we were still on the gold standard, borrowing for the dole is highly disreputable: borrowing for armaments, at any rate today is highly reputable: a strange reflection perhaps on our civilization, but true."

He stressed that before such borrowing was attempted, "it should be convincingly shown that taxation (already exceedingly onerous) has been carried to the highest point that the country can bear without grave detriment to the economy." If the Ideal Scheme were accepted, he suggested that the amount raised from revenue for defence for 1936 should be £170 million, and that that amount should rise to £215 million by 1941 if the Government intended to stay within the realm of sound finance.[50] In short it would mean tax increases.

Warren Fisher "entirely share[d] Sir Richard's dislike for a Rake's progress: and the facile and mischievous nonsense preached by distinguished economists since the war provides an excellent argument for caution." He also agreed that " 'sound finance' includes some regard for expediency, and is something short of the absolute law and prophets." When it came to an immediate increase in taxation to go along with the borrowing necessary for the Ideal Scheme, however, he was concerned about the impact on public opinion.

> So little are they conscious, in my opinion, of the dangers which are looming everywhere that the one hope is to educate them in every way. This will need time, but if the first stage of the process takes the form of the tax collector ramming his hand still further into their pockets, may not the result of such "sound finance" be that we shall be forced to continue to go unarmed?

He feared that such a tax increase would result in a taxpayers' revolt that would prevent rearmament and "inure [sic] solely to the benefit of Germany in the form of savage impositions by her."[51]

THE 1936 BUDGET

The question of whether to tax or not to tax quickly moved from the realm of theory to that of practice when the 1936 budget was formulated. In spite of the fact that revenue had increased at a greater rate in 1935 than had been anticipated, the rate of increase of defence expenditure had expanded even more rapidly, leaving

[50] T 171/324, 12/2/35, Hopkins on the budgetary effects of the Ideal Scheme.
[51] T 171/324, 12/2/35, Fisher on Hopkins' note.

the Treasury with the prospect of a £20 million deficit in 1936 if taxes were not raised.[52] This increase in the defence estimates had nothing to do with the Ideal Scheme, but was mainly the result of the new R.A.F. program that had been authorized in 1935. As the time for drawing up the final proposals for obtaining revenue drew near, the Treasury found itself in a quandary. On the one hand the imperatives of orthodox finance dictated that the funds be raised through a tax increase, while on the other, there was concern "that the effect [of such an increase] on trade and industry will be severe and economic recovery now so promising will finally be checked."[53] In the event, the Treasury decided to produce the amount by increasing the income tax by 3d (pence) in the pound (1.25 percent after exemptions and deductions) and raising some indirect taxes.[54] This was the first budget on which rearmament had an obvious impact. The Treasury was uneasy about the effect it would have on economic recovery, and by the same token had no illusions about the repercussions of the Ideal Scheme if it were ever authorized.

The Treasury position on the Ideal Scheme, as set forth by Chamberlain in the course of the D.P.R.'s consideration of it, was that in its entirety it would exceed Britain's capacity to pay for it without dislocating the national economy. Therefore it would have to be scaled down to the point where it would be in line with the nation's resources; and in scaling it down, priorities would have to be set in order to put those resources to optimal use. Chamberlain argued that an R.A.F. with a powerful striking capacity offered the greatest security for the amount available to be spent. The Navy would receive the second priority, and the Army a distant third under these criteria. This line of argument was in effect the same one used in the discussions of the first D.R.C. report. Deterrence in this case, however, was no longer supposed to discourage Germany from creating an air force (that theory had failed), but was supposed to discourage her from considering an attack on Britain or on one of Britain's allies. This theory was the

[52] T 171/324, 1/9/36, Hopkins to Chamberlain on the budget.

[53] T 171/324, 2/24/36, Hopkins memorandum to Chamberlain on the 1936 budget.

[54] B.E.V. Sabine, *British Budgets in Peace and War: 1932–1945* (London: Allen and Unwin, 1970), p. 80.

precursor of the theory of nuclear deterrence that held sway in the 1950s, and it shared its flaws, the most important of which was its inability to deal with limited provocation. As with the first D.R.C. report, Chamberlain found himself opposed by Maurice Hankey and Stanley Baldwin, who felt that a balanced approach to rearmament would provide the optimal security. It should be added that neither Hankey nor Baldwin advocated spending on the scale recommended in the Ideal Scheme, but they believed that the amounts allotted to the services should be balanced. The difference in opinion between the two most powerful politicians in the Government resulted in a stalemate that was broken only when Baldwin stepped down and Chamberlain became Prime Minister. The bone of contention in this case was not the Navy and the Far East, for the Navy, as we shall see, was granted the building program it had been fighting for, but the Army and the importance of the Field Force.

DECISIONS ON DEFENCE

The plight of the British Army in 1936 was an unhappy one. Operating on a subsistence budget as the result of the Treasury's successful insistence that their projected allocation in the first D.R.C. report be cut in half, they suffered a chronic inability to attract recruits, and faced the impossible task of preparing a force to operate on the Continent without men, money, and, many said, imagination as well. The only rays of hope that shone on this dreary scene were the recommendation in the D.R.C.'s Ideal Scheme that the Army be adequately funded to fully equip a Field Force for operation on the Continent as soon as possible, and the appointment of Alfred Duff Cooper as Secretary of State for War after the November elections. Duff Cooper proved to be a tireless and passionately outspoken fighter for funds for his service and a man who did not hesitate to incur the wrath of the powerful in the Cabinet if he felt it would serve this end.

The opposition he faced in the defence requirements subcommittee of the D.P.R., which was formed to consider the Ideal Scheme, was formidable. Headed by Chamberlain, whose opinions on the utility of the Army had only hardened since the discussion of the first D.R.C. report, that opposition was given added substance by Lord Weir, who was the subcommittee's technical expert on industrial affairs. Chamberlain's view was that the nation had limited resources, and that they "will be more profitably employed in the air, and on the sea, than in building up great armies."[55] Lord Weir's insights were more specific and technical, hence more damaging. He contended that the proposals in the Ideal Scheme for the Army to reach a state of readiness by 1938–39 could not be realized, even given virtually unlimited funding, because the industrial capacity to manufacture the equipment could not be created in that amount of time. Moreover an effort to do so, besides being enormously costly, would require vast amounts of skilled labor that would have to be drained from the civilian sector, and would create demands that the machine tool industry probably could not meet. The upshot would be a massive dislocation of the industrial economy. He argued that to build an air force on a large scale would cause less dislocation and cost less because there already was an aircraft industry with plant, equipment, and a skilled labor force, the expansion of which could be carried out far more efficiently.[56] When these contentions were added to the argument that an effective air force would have a greater impact on an invading enemy army than the best small mechanized army money could buy—an opinion that was not only the Air Force's but that of the eminent military critic, Basil Liddell Hart[57]—the case against the Army appears quite overwhelming. It was argued within the basic context set forth by the Treasury that the nation had limited resources for defence, and therefore had to establish priorities to insure their optimal use.

Duff Cooper and Hankey responded by making three points, the most telling of which was that unless Britain had an effective force to send to the Continent her allies would feel she was not making a significant contribution to their alliance and would act accordingly, a point emphasized in the earlier Army debate. The second point was that there were many situations in which a large air force with all its capacity to wreak havoc could not be used,

[55] Feiling, *Chamberlain*, p. 314.

[56] CAB 16/112, D.R.C., 28th meeting, 10/2/35.

[57] Peter Dennis, *Decision by Default: Peacetime Conscription and British Defence, 1919–39* (Durham, N.C.: Duke University Press, 1972), pp. 62–65.

but a small, effective army would have great impact. This idea, spoken of today as flexibility of response, was the rationale behind the idea of balanced forces. The third point was simply that if the spending were not done to expand the nation's war potential in terms of industrial capacity, it was estimated that it would take the Army fifteen to twenty years to reach the level of readiness considered necessary by 1938–39.[58]

Stanley Baldwin found the arguments on both sides compelling. He was especially aware of the political as well as the economic repercussions of equipping a large expeditionary force, aware, that is, of the extreme unpopularity among the populace of committing British soldiers to fight and die defending the borders of foreign nations. The strategic imperatives were not lost on him either, and his solution to the conflict was a characteristic compromise. He suggested that a Field Force of four infantry and one mechanized divisions be equipped for intervention on the Continent within five years, but that the preparation of the divisions of the Territorial Army, the T. A., for their role as the reserve force be put off. This solution made neither side happy, but, with Baldwin behind it, it was the best either could expect to obtain. The Cabinet authorized the proposal in early February, postponing the equipping of the Territorial Army until such time as the industrial situation would permit.

Within days of the Cabinet's decision, the German Army moved into the demilitarized zone of the Rhineland, as Europe watched in confused surprise. The French were unprepared to confront the Germans, and the British, for all their protestations about the Germans walking into their own back yard, were militarily unequipped to try the issue even if it had been vital to their national security. The C.O.S. reported that Britain could only have sent in a symbolic force of five brigades to "show the flag," and that even these would not have been fully equipped for war. The C.O.S. went on to say that "if war were to break out, we should not be able . . . to mobilize a force with which to reinforce France or Belgium on land for a considerable time."[59] Here

[58] CAB 16/112, D.R.C., 28th meeting, 10/2/35; CAB 16/123, D.P.R. (DR), 2nd meeting, 1/14/36.
[59] CAB 24/261, CP 81(36), 3/17/36, C.O.S. on the Rhineland crisis.

was a graphic illustration of the points that Hankey and Duff Cooper had been making, but to no avail. The decision had been taken.

During the same period that the Army was fighting for the survival of the Field Force, the Admiralty was busy trying to win approval for the most ambitious recommendation in the Ideal Scheme, the two-power standard. This represented a considerable increase over the seventy-cruiser fleet, for which they had striven for three years without success, and was the most expensive of the proposals in the Ideal Scheme. The Navy based its request on the failure of the recent Anglo-Japanese-American naval discussions, and the newly negotiated naval agreement with Germany that allowed her to build her surface fleet up to 35 percent of Britain's fleet strength.[60] Such was the stringency of the budget, the Treasury had to oppose the recommendations. Their very scope, however, made the seventy-cruiser program appear inexpensive by comparison, a result the Navy no doubt desired. In any event, the Treasury reluctantly gave the seventy-cruiser program its blessing, and on February 25, 1936, the Cabinet approved it, authorizing construction to begin. The approval was a victory, but the Admiralty considered it the least that it deserved. In January Eyres-Monsell had made a determined effort to lay before the Cabinet naval estimates for the upcoming year based on the two-power standard. The Treasury opposed such action on the grounds that the D.P.R. had not had time to evaluate the recommendations in the Ideal Scheme, and that no action could be taken on those recommendations before it had. Eyres-Monsell insisted that it was vital to the security of the nation that the decision be taken immediately, but Baldwin refused to go along. In supporting the Treasury he wrote to Monsell that "though it appears to be only a point of procedure, it raises considerations of finance—in fact the financing of the first stage of the expansion scheme."[61] Eyres-Monsell of course complied with the Prime Minister's wishes, and completed the Admiralty estimates on the basis of the existing Cabinet authorization. After the Rhineland crisis, and the deterioration of relations with Italy, which necessitated the mobilization of the fleet in the Mediterranean, the Navy requested and received

[60] Bright, "Britain's Search for Security," pp. 345–374.
[61] P.R.E.M. 1/200, 1/16/36, Baldwin to Eyres-Monsell.

a supplementary increase in their budget. The Treasury suspected that "the Admty [sic] are deliberately accelerating the modernization of equipment, no doubt thinking (as on the last occasions) that now is the time to get what they want."[62] They did not get the two-power standard, although the Cabinet did accept the need for it in principle.

The Ideal Scheme could offer nothing to the R.A.F. that it had not already been given. In fact it was a threat in that it implied that the nation's resources should be distributed more evenly among the services rather than being disproportionately divided in favor of the R.A.F. and the Admiralty. Although it had been authorized to undertake a large expansion program in May of 1935, the R.A.F. prepared expensive new suggestions for the Ideal Scheme lest the Air Force lose the ground it had gained relative to the two other services. The proposals would have added £75 million to the Air Ministry budget over the coming five years, and concentrated on the creation of reserves, as well as on the build-up of supplies of jigs and gauges needed to assure the nation's productive capacity in time of war.[63] In the event, nothing came of the R.A.F. proposals in the Ideal Scheme. It already had authorization to buy more airplanes than the industry could build, and was engaged in trying to encourage the aircraft industry to create more capacity for production. It is safe to say that in 1936 money was not an object for the R.A.F.

By late February the D.P.R. had decided which elements of the Ideal Scheme would be incorporated into the expanded rearmament program, and the service estimates had been authorized by the Cabinet on the basis of the D.P.R.'s decisions. A second White Paper on defence was then prepared to inform the nation of the Government's plans. Presented in early March, the White Paper cited the threat posed to Britain's security by the rapidly expanding military forces of other nations, and described in general terms the measures the Government intended to take to bring the nation's defences up to standard. It warned that if steps were not taken immediately, Britain would find herself in a position of permanent vulnerability. The Government hoped the scope of the

[62] T 161/718/40836, 4/17/36, Alan Barlow to Hopkins on the Admiralty supplementaries.
[63] CAB 16/112, D.R.C. 30, 10/2/35, Chief of the Air Staff Ellington on the additions to the R.A.F. program.

proposed program would convey to the British people the serious-ness of the situation and unite them behind the plans to rectify it. Intended to quiet the criticism of those who claimed that the Gov-ernment rearmed too much as well as that of those who held that it did too little, the White Paper satisfied neither group. Both sides castigated it for its vagueness, pointing out that it gave no specifics on how much was to be spent or on what. Conservative critics felt it further demonstrated the inadequacy of the Govern-ment's planning, while Labour saw it as further evidence that the Government was intent on full scale unilateral rearmament with-out regard to collective security or the League of Nations. *The Economist* condemned it as

> a confession not of faith but of failure . . . a tragic admission by the "old guard"—Mr. Baldwin, Mr. Ramsay MacDonald, the Service Chiefs and the rest—that the task of leading the world along the path of sanity has proved too much for them; and that they have fallen back, in something akin to panic, on the fatally familiar shifts of the pre-war world.[64]

The White Paper changed few minds and sparked no sense of na-tional urgency. Rather than quiet the critics, it provided them with additional fodder for their attacks.

LABOUR'S ATTACK ON THE BUDGET

The presentation of the 1936 budget to Parliament presented the Opposition with its next opportunity to challenge the rearmament program, and it took full advantage of it. As noted earlier the Government was forced to find £20 million in additional revenue in order to cover the increased cost of rearmament, and chose to increase taxation to raise it. Labour questioned the need for arms expenditure on the scale the Government proposed, expressed its doubts about the sincerity of the Government's efforts to prevent profiteering, and scored the Government's intention to raise part of the needed revenue through an increase in the duty on tea. Labour MP Arthur Greenwood, in a speech opposing the increase in defence spending, noted that:

> Every penny of expenditure on armaments for destructive and non-constructive services helps to twist the economic life of the

[64] "The Dangers of Defence," *The Economist*, 3/7/36, pp. 511–512.

nation out of its normal course. . . . What is going to be the effect of all this expenditure on armaments, when the money has been spent? Social wreckage again and again, in area after area, in industry after industry.[65]

The most passionate debate centered on the Government's intent to increase the duty on tea by 2 pence per pound, an indirect tax that would weigh most heavily on the poor. The issue of direct versus indirect taxation had come up almost annually at budget time since the National Government had come to power. Labour stated quite flatly that "it is the deliberate policy of all Tory Governments to increase indirect taxation and decrease direct taxation. It is euphemistically described as 'broadening the basis of taxation.' "[66] *The Times*, in an editorial reply to a Labour attack on indirect taxation, stated the conservative view in these terms:

It would be better for the Labour Party, . . . if it really has the increased prosperity of the nation at heart, to join in advocating the reduction of taxes which fall unquestionably upon British industry, instead of trying to transfer into our financial system their untenable political distinctions between "haves" and "have nots."[67]

Greenwood pointed out that the tax in question, the tea duty, would cost a poor family with an annual income of £100 at least 7 shillings a year which "means two pairs of children's boots per year, it may mean a blanket, it may mean doing without something in the home." He concluded that it was "one of the meanest kinds of tax."[68] Chamberlain replied:

I have deliberately chosen the tax on tea, not because I believe it is popular but because I wanted a tax which would be widespread and which would cause as little hardship as possible. . . . In a very short time it will be forgotten, and in the rise in price due to the tax everybody will feel that they have the satisfaction of making their little contribution to the necessities of the country.[69]

[65] *Parliamentary Debates* (House of Commons), Vol. 311, 4/23/36, col. 426.
[66] Ibid., col. 422–423.
[67] Editorial, *The Times*, 5/22/35, p. 17.
[68] *Parliamentary Debates* (House of Commons), Vol. 311, 4/23/36, col. 422.
[69] Ibid., col. 436–437.

Chamberlain's was the last word on the subject. The passage of the budget itself was never in doubt as the Government majority remained intact. The issues aired and the doubts expressed, however, were not without importance. They forced the Government to face issues it would rather have forgotten.

BALDWIN'S TENUOUS LEADERSHIP

The criticism of the Government's rearmament policy by those within the Conservative party was not so easily turned aside. By May Baldwin's hold on the leadership of the party had become a topic of discussion both in political circles and the press as critics continued to question the adequacy of the plans to carry out the expansion of the services outlined in the White Paper.[70] The need for secrecy, as well as shortcomings in its organization, had forced the Government to make replies to questions raised in debate that were less than satisfactory, often appearing to verge on dissembling. This only served to fan the flames of doubt, and renewed the speculation about Baldwin's ability to provide adequate leadership for a government attempting to undertake a task of this magnitude.[71]

The speculation remained only that, however, because the Conservative critics were a diverse lot who agreed among themselves on little save the inadequacy of the rearmament program. Churchill was the most outspoken of the group that included such men as Austen Chamberlain, Sir Henry Page-Croft, Sir Robert Horne, and Leopold Amery. Even had they been able to unite in an effort to oust Baldwin, it is doubtful that they would have succeeded. Rumors to the effect that they were considering trying to unite with Labour critics of the defence programs to form a "popular front" were also obviously off the mark as the two groups were diametrically opposed in their views on the course rearmament should take.[72] Nonetheless, the very existence of

[70] On July 17, 1937, Leslie Hore-Belisha, then Minister for Transportation, frankly acknowledged the decline in Baldwin's prestige and effectiveness during the period after the election in an off-the-record interview with W. P. Crozier, the editor of the *Manchester Guardian*. W. P. Crozier, *Off the Record: Political Interviews, 1933–1943*, ed. by A.J.P. Taylor (London: Hutchinson, 1973), p. 62.

[71] "Mr. Baldwin and the Rebels," *The Economist*, 5/30/36, p. 479.

[72] "His Majesty's Government," *The Economist*, 5/16/36, pp. 349-350; see also Middlemas and Barnes, *Baldwin*, p. 941 and p. 945.

such rumors and intrigues, unrealistic and farfetched as they might have been, demonstrated both the failure of the Government to unify even its own followers behind the defence program, and the continued deterioration of Baldwin's political standing.

Had it not been for Neville Chamberlain's presence in the Government and his loyalty to Baldwin, those intrigues might have had some chance of success. As it was, however, although some of the Government's followers' confidence in Baldwin might have been shaken, the assurance that Chamberlain, his acknowledged successor, would soon be Prime Minister kept them from looking further for leadership. Baldwin, who earlier had expressed his wish to retire after seeing the Government through the 1935 election, was known to be seeking an opportune moment to step down. Most Conservatives were confident that when he turned the premiership over to Chamberlain, the Government would have the leadership it needed.

The problem was that the opportune moment turned out to be much longer in coming than anyone had anticipated. Baldwin stayed on through the spring in order to see that the rearmament program was well under way and to avoid the appearance of bowing out under political pressure. In June, his health, which had not been robust, began to fail, and by late July he was suffering from complete nervous and physical exhaustion. His doctor ordered him to drop his work entirely and seek complete rest. Not wishing to end his career on this note, he turned most of his day-to-day administrative and decision-making responsibilities over to Chamberlain in early August, while he left London for the country in order to regain his health. When he returned to 10 Downing Street in October, he still lacked sufficient stamina to resume all of his duties. As a result Chamberlain was left in charge of the Prime Minister's day-to-day affairs, while Baldwin used what strength he had to see England through the abdication crisis. It was not until June 1937 that Chamberlain became Prime Minister in his own right.[73]

ITS EFFECT ON REARMAMENT

Baldwin's lingering premiership placed Chamberlain in an

[73] Middlemas and Barnes, *Baldwin*; for more on Baldwin's health see pp. 960-973.

anomalous position and had certain detrimental effects on the rearmament program. While Chamberlain was clearly the dominant force in the Government, Baldwin retained the ultimate position and the power that went with it, making independent policy initiatives on Chamberlain's part a doubtful proposition. The Cabinet remained Baldwin's Cabinet, and any such initiatives would have been fraught with political complications. Chamberlain, therefore, felt constrained to act as a caretaker, biding his time until the reins of power were firmly in his hands before initiating any new departures.

During this period the rearmament program suffered from a lack of direction. Baldwin had never developed an overall concept of the shape the rearmament program should take, preferring instead to let it evolve, and to use his influence to guide that evolution at critical moments as he felt circumstances required. His actual contribution to policy was, therefore, limited to the situations in which he intervened. The decisions on defence taken in January and February 1936 followed this pattern, being rather general in nature and suiting the circumstances of the time. By the second half of 1936 new circumstances had arisen: the services, in competition with one another, were setting forth all manner of expensive new programs, and the Cabinet lacked either leadership or a clearly delineated policy to assist in making choices between them. Chamberlain had definite opinions on many of the issues that arose, but if they involved new policy departures or the reversal of policies that Baldwin had helped to determine, he refrained from imposing them.[74] As a consequence supplementary programs were authorized by the Cabinet for no reason other than that the sponsoring services had made them appear vital. The Treasury tried to hold the line on the burgeoning expense, but it was not until after Chamberlain became Prime Minister that an effort was made to develop a unified defence policy.

It is clear that Britain's rearmament program got off to a poor start in 1936. The Hoare-Laval affair robbed Baldwin and the

[74] The question of the role of the Army provides a good case in point. Although Chamberlain differed strongly with the compromise decision imposed by Baldwin, he refrained from trying to force the Cabinet to reverse that decision until he was firmly in control.

Government of the initiative and political credibility they had gained in their election victory, initiative and credibility they had hoped to use to win the nation's support for their rearmament policies. The Ministry of Defence debate then served to put them on the defensive, revealing the splits within their own party on the issue, and destroying any chance there may have been of achieving unified support for their programs. The unwillingness of Baldwin and his Cabinet to make final decisions on such matters as the Admiralty's demands for the New Standard and the Army's requests for the T. A. contributed to the shortcomings of the defence program by failing to give it direction. Purpose was sacrificed to flexibility, and as a consequence government policy lacked cohesion. Baldwin's illness and subsequent lingering premiership served to perpetuate the vacuum in leadership until he finally stepped down. Much of the responsibility for the failures of this year must rest with Baldwin, who was incapable of providing the necessary leadership himself yet refused to step aside when his health began to fail, for someone who could. Churchill summed the situation up well, in the course of a speech condemning the deficiencies of the Government's defence program, when he told the House of Commons on November 12, 1936:

> The Government simply cannot make up their minds, or they cannot get the Prime Minister to make up his mind. So they go on in strange paradox, decided only to be undecided, resolved to be irresolute, adamant for drift, solid for fluidity, all-powerful to be impotent. So we go on preparing more months and years—precious, perhaps vital to the greatness of Britain—for the locusts to eat.[75]

THE UPSHOT OF THE IDEAL SCHEME

The implications of the rearmament program at the end of its first year of active endeavor were manifold. In the short term the foreign policy to be pursued was clearly stated by the D.R.C. in the introduction to the Ideal Scheme:

[75] *Parliamentary Debates* (House of Commons), Vol. 317, 11/12/36, col. 1107.

> We consider it to be a cardinal requirement of our national and imperial security that our foreign policy should be so conducted as to avoid the possible development of a situation in which we might be confronted simultaneously with the hostility, open or veiled, of Japan in the Far East, Germany in the West, and any Power on the main line of communication between the two.[76]

The reference to the power on the main line of communication between East and West meant Italy. Britain spent 1936 trying to get back on good terms with her, with only mixed results. The British were keenly aware that they were operating from a position of weakness, and the hope expressed in the Ideal Scheme was that she could stay on good terms with the threatening powers until she had rearmed sufficiently to be able to deal with them from a position of strength.

How Britain should prepare to deal with these powers was a point of some contention. The Treasury view, which prevailed, was that the R.A.F. should be built, in Chamberlain's words, into a force "of such striking power that no one will care to run risks with it."[77] The emergence of the R.A.F. as the centerpiece of British rearmament was by no means based on simply strategic considerations. They appear in fact to have been of secondary importance. Politically the considerations that disposed the Government to prefer to build up the Air Force rather than the Army—for that was always the question after it was agreed that Germany was the threat to be prepared against—were two. First was the public concern and fear about the effects of bombing and the threat of the "knockout" blow. It was a fear based, as we have seen, on the experience of the First World War, built up by the Air Ministry, and communicated to the public by government ministers whose own concerns had been aroused. By 1935 the public concern was considerable and was a factor in the decision to proceed with the expansion of the R.A.F. The public's antipathy to becoming involved in a war on land on the Continent, although this never became a public issue, was known to the ministers, who had no desire to make it one. Thus the political considerations favored the emphasis on the R.A.F. The major considera-

[76] CAB 16/139, D.P.R. 52, 11/21/35, the Ideal Scheme.
[77] Feiling, *Chamberlain*, p. 314.

tions, however, were economic. The fact that the expansion of the Air Force would be less costly, cause less dislocation of the economy, and generally provide more deterrence value for the money spent made it infinitely more attractive to the Treasury than the expansion of the Army. Because the Treasury was the single most powerful ministry in the Government, the emphasis on the Air Force as a matter of policy followed naturally from the Treasury's advocacy. It was an advocacy that was to become less wholehearted as the R.A.F. program became more expensive.

The implication of the policy of deterrence in terms of foreign policy was isolation from the Continent and to a lesser degree from the Empire. The reluctance to create an Army that could effectively intervene on the Continent left Britain with only the bombing deterrent to influence Germany, and although it might deter her from attacking Britain, it was not a force that could influence Germany's behavior toward other states. While the Treasury did relent and permit the Navy to begin to build up to the seventy-cruiser level, it was recognized that anything less than the two-power standard would, in the event of simultaneous conflict with Germany and Japan, leave the Empire in the Far East with a less than adequate naval cover against the Japanese. In June of 1934 Neville Chambelain already had set forth the idea that was to become the basis of British defence policy when he wrote:

> My first proposition . . . is that during the ensuing five years our efforts must be chiefly concentrated upon measures designed for the defence of these Islands.[78]

[78] CAB 16/111, D.C.M. (32) 120, 6/20/34, Chamberlain note on the finance of defence.

CHAPTER III

Industrial Mobilization for Rearmament

WAR POTENTIAL

WHEN the suggestion began to be heard with increasing frequency, in the offices and corridors of the service ministries along Whitehall in late autumn of 1933, that the new Government might be moved by the C.O.S.'s rather alarming annual strategic review of the world situation to at last take some steps to remedy the deficiencies that had been allowed to accumulate in the nation's armed services, a small group of obscure men took special notice. They were the members of the Supply Board, the central committee of the Principal Supply Officers Sub-Committee of the C.I.D., and it was to be their responsibility to arrange for the actual procurement of much of the materiel that the services would be seeking when and if the Government agreed to their programs. The Supply Board was naturally concerned with the capacity of the nation to produce the goods it was called on to procure, and in the autumn of 1933 it was acutely aware of the shortage of that capacity. To deal with the situation, the members decided to invite three of the nation's leading industrialists to advise the board on how to go about resurrecting Britain's war potential, as the industrial capacity to produce war-related goods was called. The industrialists were Lord Weir, who later joined Swinton in a similar capacity at the Air Ministry, Sir James Lithgow, the head of a leading Scottish shipbuilding firm, and Sir Arthur Balfour, a leading steel manufacturer. Lieutenant-General Charles gave the three men a very succinct description of the supply situation.

He pointed out that without the whole-hearted co-operation of industry it was quite impossible to carry on a war to a successful conclusion. During the lifetime of those present, i.e., in South Africa and again in 1914, avoidable and needless expenditure of both lives and money had occurred due to the failure to supply our forces with adequate arms and munitions. The

Organisation was engaged in locating the capacity of normal peace-time industry to turn over to the production of war stores. Until such time as supply could overtake expenditure in the field, reliance must be placed on reserves and the products of Government factories and existing armaments firms. Investigations hitherto made had revealed that there was likely to be great difficulty and delay as all war-time plant had been scrapped, and the armaments firms had virtually ceased to make munitions. As this was a manufacturing problem, the advice of Industrialists was sought as to the best means of solution.[1]

Lord Weir, who had been an adviser to the Ministry of Munitions in the First World War, felt the plans to locate plant that could be turned over to wartime production were fine, but suggested that capacity for the production of some highly technical goods should be available in the event war broke out. He proposed that large industries doing technically related work might be persuaded to set up such plant at government expense, to be put into use at the outbreak of war. This suggestion was the precursor of what came to be known as the "shadow factories" scheme. Weir also stressed that "approaches should only be made to the big firms and the big men." The industrialists agreed that closer contact with industry was a matter of extreme importance, and that, in view of Russia's and Germany's capacity to produce arms in peacetime, Britain had to build up her capacity as well. In April of 1934 Weir presented a paper in which he stated that it was "essential to create a shadow armament industry capable of expansion to meet war requirements," necessary to expand existing government and private arms plant, and important to complete the industrial census in order to locate the 250–400 major firms best suited to convert to arms production if war should break out.[2] This paper was passed on to the C.I.D., which approved its recommendations in principle.[3] That, however, was as far as the Government was willing to commit itself until the German air parity scare of

[1] SO 241/Supply 3/5, P.S.O. 407, 12/19/33.
[2] SO 241/Supply 3/5, P.S.O. 421, 4/26/34.
[3] CAB 2/6, 264 (6), 5/31/34; T 175/47, 5/30/34, Treasury memorandum on Weir paper.

1935. The reasons were the same as those that limited the financial expenditure on rearmament, of which industrial expansion was an integral part.

The German scare resulted in Weir's not only becoming Swinton's top adviser at the Air Ministry, which was to oversee the most rapidly expanding sector of the nation's industry in the coming years, but the industrial adviser to the D.R.C. and the D.P.R. In a letter to Baldwin, Weir expressed his opinion on the direction the mobilization of industry to meet the German threat should not take. He wrote that he "was averse to doing anything which would turn industry upside down by creating a war spirit and practice, but [he] felt that we must quietly but very rapidly find an effective British compromise solution as opposed to merely copying the centralised dictator system."[4] Three months later in a letter to Swinton he asked:

> Are we doing all we ought to to anticipate by proper planning and arrangements the grave delays which were the feature of our almost fatal unpreparedness in 1914? More than that, and apart from mere paper planning, have we anticipated our definite weaknesses in our facilities for producing war materiel which undoubtedly exist, and are we taking rapid steps to strengthen these by definite and corrective action? It is my clear impression that we have not done and are not doing this.[5]

His clear impression was of course correct, but the situation was beginning to improve, due in no small part to his efforts.

GOVERNMENT, INDUSTRY, AND REARMAMENT

The Government's rearmament program became a source of concern to industry in 1935 when it became clear that the program was going to have an effect on the economic and industrial structure of the United Kingdom. The Federation of British Industries, F.B.I., the trade organization that represented most of the major British industrial firms, requested a meeting with the Government to discuss the future of the relationship between government and

[4] Bald. #1, 5/18/35, Weir to Baldwin.
[5] T 172/1830, 8/22/35, Weir to Swinton on the air rearmament program.

industry. The meeting took place on October 17, with the Government represented by Baldwin, Chamberlain, and Sir Horace Wilson, who was the Chief Industrial Adviser to the Prime Minister. Representing the F.B.I. were five of the nation's leading industrialists, Sir Francis Joseph, Lord Herbert Scott, Sir George Macdonogh, Sir George Beharrell, and Mr. Guy Locock. Sir Francis Joseph, the spokesman for the delegation, began by explaining that the federation had decided to ask for the meeting as the result of a resolution moved at the Bournemouth meeting of the Conservative Party, which was drawing up a platform for the upcoming election. The resolution, which had "caused some concern to industrialists," suggested that the Government consider " 'organising industry for speedy conversion to purposes of defence if need be.' " Sir Francis continued:

> They [the industrialists] appreciated the Government's anxiety to take all necessary measures for defence and the Federation of British Industries were ready to co-operate with the Government to the fullest possible extent. . . . They realised the concern that might be felt about the cost of any defence programme and they were anxious to co-operate to prevent exploitation in any practicable way, but they would like to know what might be required and what was implied by the words of the Resolution.

At this point the reason for the industrialists' concern was laid bare.

> They felt, however, that industry ought to provide any organisation which was necessary from within and not have it imposed upon them from without. What form of organisation was envisaged? If it was organisation within and by industry itself they would be ready to co-operate in every way.

Implicit in this statement is the rejection of any effort by the Government to compel industry to organize for defence production through the force of law. This reaction was prompted in part by the memories of "the waste and extravagance that had arisen in the last War," which to their minds had resulted from the Government's control of wages, prices, profits, and production. The experience had reinforced their belief in the efficiency of the market system and their natural disinclination to allow any interference with the way they chose to employ their capital.

Chamberlain proceeded to put their minds at ease on that score. He stated:

> The Government had no idea of nationalising or socialising industry or of imposing an organisation from without. They contemplated, however, that the programme would be one of considerable magnitude and considerable organisation would be required to prevent overlapping and competition in demands for skilled labour, and to reach an understanding about priorities. . . . The time would come when the Government would be in a position to talk more in detail to industrialists, but they wished that as far as possible industry should carry out the necessary organisation itself.

Chamberlain then took the opportunity to air some Government complaints and utter some cautionary words to the industrialists on the subject of self-control by industry. He began by pointing out the dangers of allowing prices to rise: rising prices would lead to demands for wage increases, which in turn would lead to more price increases and quickly into an inflationary spiral. "This would have two unfortunate effects: (1) the Government would have to pay too much for its requirements, and (2) these rises would not be confined to munitions, etc., but would spread to industry generally and competitive industry would be faced with grave problems arising from higher wages and higher costs." Having established these effects, he told the industrialists that in the two industries (steel and textiles) to which the Government had given the right to organize themselves, and to which it had granted special tariff protection in the hope that they would make themselves more efficient, there had been just such rises in price. What was worse, "there had been no satisfactory explanation of the rise in prices, and there had been a refusal to allow the Contract Departments to examine costs." In such a situation there was a danger "lest there should be general accusations of profiteering." To prevent such accusations "it was most desirable that the Government should be in a position to say that it had been met with great candour and that there had been willingness on the part of industrialists to show their costs."

The Government's message was summed up by the Prime Minister, who said in conclusion:

> The Government had no wish to impose an organisation upon industry, and if industrialists would play the game there would be no need to do so. It was in their interests to play the game both as industrialists and as tax payers. He felt sure that they would agree with him in feeling that the dangers to which the Chancellor had called attention must be avoided.[6]

In a letter to the Government following this meeting, the F.B.I. reiterated its assurance of cooperation, and stressed that the nation's industry could supply the Government's needs at a reasonable cost.

> It is the considered opinion of the Federation that these objects can be achieved most efficiently and most economically if the industries of the country are taken into the confidence of the Government and charged with the task of themselves setting up any organisation required to meet Government needs and ensuring that the work is carried out on a reasonable basis of cost both to the industry and to the Government.[7]

The significance of the industrialists' message to the Government should not be understated. They implied in a thinly veiled threat that if the Government were to attempt to use any form of compulsion in time of peace to order the production of war materiel industry would not cooperate. There was no attempt to spell out what non-cooperation meant. It was understood that it would result in the flight of capital, lockouts, and the widespread economic and industrial dislocation that always occurs when the spectre of socialism casts its shadow on a free enterprise economy. Compulsion was seen as the first step on the road to socialism, and it mattered little who implemented it. At the same time industry was reminding the Government of the interest they shared in continued social, economic, and political tranquility. They were, after all, on the same side of the political fence.

---

[6] CAB 16/112, D.R.C. 38, 10/17/35, minutes of a meeting between Baldwin, Chamberlain, and a delegation from the F.B.I.
[7] CAB 16/123, D.P.R. (D.R.) 7, 1/8/36, letter from the F.B.I.

THE DOCTRINE OF COOPERATION

The Government needed no such reminders. It had already laid its general plans along the lines suggested by the F.B.I. At the heart of both industry's proposals and the Government's plans was the understanding that they would work together. The guiding ethos was what may be called the doctrine of cooperation, which Stanley Baldwin described in a speech to the Commons on March 9, 1936. In his words:

> We are proceeding in confidence that we shall get the goodwill of the industry. Our plans assume contact and collaboration between the Government Departments and industry right up the scale with firms, branches of industry, and overhead organisations. We feel sure that experience will teach us that we must rely on safeguards combined with the co-operation of industry to perform a national service without the making of undue profits. We have no dictatorial powers. We cannot make our preparations in secret nor can we in time of peace force anyone to participate in them. But we can count on goodwill in a way and to an extent not available to a dictator, and we hope and believe that we shall secure it.
>
> . . . .
>
> It is the Government's part so to plan their programme as to fit in with practical considerations, and generally to co-operate with industry in its execution. On this footing, and with a sufficient degree of goodwill among all concerned, I do not see why the programme should not be carried through without any great dislocation of our industrial economy, and certainly without any menace to organised labour or to trade union standards.[8]

The goodwill that Baldwin kept referring to was the business term for that intangible attitude or atmosphere that was vital to smooth and efficient cooperation. The essence of goodwill, to the degree that it can be defined, is a. trust on the part of each party that the other is doing its utmost, and b. a willingness not to quibble over pecuniary differences. Good will is the basis of efficient

[8] T 161/718/40733, 3/25/36, from Hansard, 3/9/36, col. 1843–1844.

business relationships, and it works because both parties stand to gain from it. As the basis of the relationship between government and industry it leaves something to be desired. That relationship is in reality more like one between a producer and a consumer; and when it comes to buying arms, a consumer facing a cartel at that. Thus the trust required is nearly absolute, because the government is almost wholly reliant on the assistance of the producer in checking its costs. Suffice it to say, the way of cooperation between the Government and industry was not always smooth, despite their shared interests.

INDUSTRIAL MOBILIZATION

After the election had been safely won, the Government turned to preparing its plans for the mobilization of industry to deal with the critical shortages in the nation's war potential. The overall idea was to "build up in peace-time reserves sufficient for a limited period after the outbreak of war, and simultaneously plan and arrange our industrial capacity in peace-time so that in the interval assured by the reserves it is able to turn over to full war production." This build-up was to be executed "without interference with or reduction of production for civil and export trade. From the production point of view this greatly complicates the matter, but any such interference would adversely affect the general prosperity of the country and so reduce our capacity to find necessary funds for Service programmes."[9] The stipulation that the rearmament program would not interfere with civil or export trade was authorized as policy by the Cabinet and made public in February. It was a concession to the concerns of the industrial community about Government interference with business.

The Government took a number of steps to accommodate its bureaucratic structures and procedures to the needs of its contractors. Under the existing procedures each of the services had its own contracting department that, after receiving authorization from the Cabinet to purchase a given item, would put the contract for the production of that item out for competitive bidding or, as was increasingly the case, negotiate the terms of the contract with

[9] CAB 24/259, CP (36), 2/12/36. This was first published as a part of D.R.C. 37, the Ideal Scheme.

99

the firm best suited to produce it. The contract would then be sent to the Treasury for authorization in detail, and after it was authorized, the contractor could begin work on production. The Treasury's right and obligation to review every government contract was based on the precept that the branch in charge of finance should have the final say on all purchases contracted for by the supply branches. It was considered that the branch concerned with the procurement of funds would be most concerned to insure that the tax payers got the best value for their money. By the same token it was considered that the branch whose main concern was procuring the good should be the last to which the final settlement of price should be entrusted.[10] Although "the Government had laid down two overriding conditions to be observed by the Service Departments, (a) that the interest of private industry and trade was to be safeguarded and (b) that there was to be no profiteering,"[11] it was left to the Treasury to make the final decision as to whether a contract negotiated by a service department met those conditions. The process of passing all contracts on to the Treasury for final authorization inevitably meant a delay between the time the supply department and the contractor reached agreement and the time the contractor was actually allowed to start work. Such a situation was not exactly conducive to the swift translation of policy into action that the Government was seeking in its rearmament program. In Weir's words:

> . . . the conditions are akin in some measure to war conditions. The word of the man responsible for Supply must carry, and the spirit and enthusiasm he has evoked in the contractor's mind must not be chilled by delays of approvals, caused by financial control. I do not mean that any loose disregard should prevail on the financial side, but that the keynote must be that "the job must go ahead."[12]

The lag between the agreement on the contract with the contractor, and its final authorization by the Treasury meant that it would be that much longer before production was under way. Once production was under way, it was not unusual for the contractor to

[10] CAB 16/123, FA/OH 32, Appendix "A," 2/6/36.
[11] Weir Papers, 18/1, ID/G/167, 3/26/36, Inskip to Weir on contracts.
[12] CAB 24/259, CP 26(36), Appendix "C," 2/12/36, Weir on financial control.

find that there were unanticipated production costs, which he would ask the government contracting department to cover. The department had to forward those requests to the Treasury, and in the event of a delay in the Treasury's authorization, production often came to a halt, meaning added delay as well as added expense for the contractor. To ease this situation a committee entitled the Emergency Expenditure Committee was formed with the power to authorize expenditures up to a certain amount without further recourse to the Treasury. In early 1936 this committee, which was made up of top level Treasury and service civil servants, was renamed the Treasury Inter-Services Committee, T.I.S.C. Its creation is a good example of the Government's endeavors to minimize its interference in industry's efforts to meet the nation's defence requirements.

Other moves the Government made to encourage industry to take up defence contracts included guaranteeing that there would be a long-term demand for arms and a continuity in the letting of contracts to individual firms. The first was vital, because businessmen were very leery of becoming involved in contracts that involved "large capital commitments to satisfy demands which may not continue, and which are outside normal business requirements."[13] They remembered only too well the last war, when there was a rapid short term build-up of plant to produce arms that became totally obsolete when the war ended, because the Government abruptly stopped purchasing arms. Businessmen did not want to invest large amounts of their capital in plant that would be used only during a four- or five-year arms build-up, and would become totally obsolete when that build-up was completed. The Government's assurance was calculated to encourage investors by promising that there would be demand over a sufficiently long term to guarantee a profitable return on investment in plant. That assurance was also to a degree self-serving, because if businessmen believed that demand was going to be only for a short term, they would charge proportionately higher prices to cover the anticipated obsolescence of their plant. In other words they would write off the cost of their plant over the anticipated five-year period of demand rather than over the fifteen-year period, which

[13] CAB 16/123, D.P.R. (DR) 7, 1/8/36, letter from the F.B.I.

would be used given constancy of demand. This of course would result in a higher cost per unit produced, making the whole rearmament program considerably more expensive.

The guarantee of a continuity of orders was made "so that contractors can be attracted by business propositions, and so that personnel, and especially supervising staffs, may be trained."[14] Contractors would not be attracted by the prospect of setting up their plant to produce a good, training the labor to operate the machinery, then, upon completion of the initial contract, finding themselves waiting three months until the next contract came through. That would mean that the capital invested in the plant would not be in use over those three months, and it would also probably mean that they would have to hire and train a new group of workers, as those laid off over the three months would not likely still be available. Far more attractive was the prospect of a continuous flow of contracts that would keep their capital in constant use, thus returning a profit, and of their workforce remaining constant, thus eliminating the need for costly training of new workers. All this would be beneficial to the Government as well, because the more efficient production that would result would mean a lower final cost of the product.

To facilitate the transition to armaments work the Government also granted various kinds of subsidies to firms taking up that work. Often the subsidy was in the form of a higher price for the good produced to enable the businessman to write off his investment in plant more quickly. The Government also extended loans, often interest free, for the building of new plant, or supplied all the capital on the basis that the new plant belonged to the government, the contractor being paid to manage it. As contracts became larger, the Government began to supply more and more of the working capital, paying for contracts on an installment basis. This was especially true in the aircraft and shipbuilding industries.

CONTRACTS, COSTS, AND PROFITS

One of the early changes in contracting procedure was an increasing recourse to non-competitive contracts. These reduced the time required to let contracts, allowed the Government to assure their

---

[14] CAB 16/139, D.P.R. 56, 11/30/35. This part of the Ideal Scheme was based on P.S.O. (SB) 558.

continuity, and, most importantly, guaranteed a more rational distribution of work in industries in which demand exceeded capacity. In the latter situation companies tend to bid for work they cannot realistically handle in the belief that, should they get the contract, the Government will bail them out when it becomes clear that they are overextended rather than take the loss of the money already paid out, as well as the loss in time finding a new contractor to start over from scratch. In a situation where the industry lacked the capacity to meet demand, as was the case with the aircraft industry, it made sense for the Government to let contracts to the firms best equipped to execute them and find a way other than bidding to settle the cost.

Cost, however, was the Achilles heel of the whole doctrine of cooperation. Consult as the Government might with industries on how it could ease the transition to the production of arms, in the end the bill had to be paid, and arms did not come cheap. The Government could be accommodating on questions of procedure, but when costs were involved, it had to be able to account for and defend all expenditures. It was a matter of simple political survival. If the Opposition could prove that the Government was allowing industry to profiteer on the production of arms, the Government would not survive. When the decision was taken in May 1935 that there would be a large scale expansion of the R.A.F., Baldwin stated in Commons:

> There are two things on which I am certainly determined, and the whole Government are determined—that, in the efforts which we regard as necessary for the nation during the next couple of years, . . . there shall be no profiteering in a time that I might almost call a time of emergency. There will be a great demand for certain types of labour. There will be a great demand for the production of factories, and I hope that none of the interested parties will try to make capital out of the situation.[15]

Such assurances were easy and necessary to give, but the problem of determining the just price in a non-competitive market situation was not easy to resolve.

[15] T 161/856/49179/1, 5/22/35, Baldwin's statement to Parliament on profits and prices.

In a normal competitive market situation price and profit are regulated by the laws of supply and demand. The price of a good is a function of the demand for it and the ability of its various producers to supply it. In a competitive market situation producers vie with one another for the business of a finite number of consumers, and their competition insures that the market price of the good bears a close relationship to its production costs. The profit a producer earns is determined by the difference between the cost to him to make the good and the price at which he can sell it on the market, multiplied by the number of it he sells. In short his profit is determined by the level of demand and his own efficiency.

As long as a rough balance between supply and demand is maintained, prices and profits bear a direct relationship to cost. However, in the event of a sudden change in that balance, the relationship between price, profit, and cost will be altered to redress it. A sudden large increase in demand, which gives rise to unusual profits, is one such change. In such a circumstance the competitive market equilibrium is upset as consumers vie for goods that are scarce, driving up prices until enlarged production establishes a new balance between supply and demand. In the interim profits are unusually high. The relationship between cost and profit can be said to have broken down. The breakdown is only temporary, however, because the exceptional profits attract new capital, resulting in a growth of productive capacity. When this capacity expands to meet the new demand, the competitive market equilibrium and the resulting relationship between cost and profit are restored.

When the sudden growth in demand is caused by an urgent and massive increase in a nation's need for armaments, the market mechanism reacts as it would to any such increase. In this situation, however, the society in which the market exists condemns such behavior because it jeopardizes the society's chances for survival. The producer who insists on being paid the price the market will bear for goods vital to the national survival is condemned as a profiteer; what is normally accepted as sound business practice is considered bad form.

In Britain both industrialists and politicians condemned profiteering as a form of parasitism that was harmful to the social fabric and a menace to national security. They agreed that a free market price for armaments in such a situation was unacceptable, and rejected it. This, however, left them with the problem of determining what a socially and financially justifiable level of prices and profits should be. It was a problem that was to tax to the utmost the ethos of goodwill that the Government was trying to establish between itself and industry.

The Government had to find a rate of profit that would attract capital and management to Government work, yet not be regarded as excessive by the public. This meant that the businessman had to be offered a return that would compete with what he could expect to earn in other sectors of the economy. While the requisite and acceptable levels of remuneration could be generally estimated, the conditions peculiar to this situation made a precise determination of the minimum return to which businessmen would respond impossible. That would be determined by the sum of their individual perceptions of the risks involved. On the positive side was the fact that, because rearmament was vital to the nation's security, the Government was certain to see it through. This meant that business was assured, at least for the duration of the arms build-up, and, more importantly, that ample government capital would be available. On the negative side was the uncertainty of government demand in the middle and long term. If after the arms build-up the world situation took a turn for the better, the businessman faced the prospect of being left with a large capacity to produce arms and no one to sell them to. The major imponderable, of course, was what kind of return the Government would allow to defence contractors. It may be inferred from the skyrocketing prices of aircraft industry stock that investors did not anticipate that the Government would be ungenerous. In the event, the Government's primary consideration in determining the just rate of profit was not how low a rate industry would accept, but how high a rate the public would tolerate.

CHARGES OF PROFITEERING

The Government was forced to deal with this question in Parliament when it was announced that the R.A.F. would be expanded. The Labour Opposition demanded to know what measures the

Government intended to take to guarantee that there would be no profiteering. It was an issue that was to be raised repeatedly by Labour until the outbreak of the war, and one the Government never dealt with adequately.

On May 29, 1935, two months after Baldwin had given his assurances that there would be no profiteering, Mr. Geoffrey Mander, a Liberal MP, asked if the Government was "aware that [aircraft industry] shares are being hawked round, with promises of enormous profits in a few months time: and ought not some action be taken in this regard?"[16] The question of profiteering was brought up again in July when the Government sought Parliament's approval for its plans to expand the Air Force. In the course of that debate the Opposition continued to press the Government on the steps it intended to take to insure against excessive profits by firms involved in the rearmament program. Cunliffe-Lister, the new Secretary of State for Air, who was responsible for negotiating the contracts with the aircraft industry, replied that the Government had taken on Lord Weir and other advisers, and with their assistance was negotiating contracts that would make excessive profits impossible.

Dr. Christopher Addison, the Opposition spokesman on economic matters, found this response less than adequate. As assistant to Lloyd George at the Ministry of Munitions during the First World War and subsequently as its director, Dr. Addison had learned firsthand the problems inherent in trying to prevent profiteering. Pointing out that excessive profits were possible even under the tightest contracts, he asked whether the Government intended to take any powers to compel industries involved in rearmament to keep their profits within certain limits. He cautioned that at the very least the Government required the power to inspect the books of its contractors to determine what profits were being realized, and warned that in the process of rearming Britain the Government was

> going to establish in this country a very powerful monopoly, trading not only at home but abroad, which will mean that in this deplorable business we shall have growing up a group of

[16] *Parliamentary Debates* (House of Commons), Vol. 302, 5/29/35, col. 1112.

Although, as Mr Baldwin says, air-planes offer no real protection from air attack, undoubtedly the construction of air forces stimulates industry and relieves unemployment. — *Colonel Blimp*.

2. "Gad, Sir, At Any Rate We Are Restoring Prosperity"—Low's famous Colonel Blimp congratulating an arms profiteer after it was learned in July 1935 that the Government intended to further expand the Air Force. *Evening Standard*, July 10, 1935.

powerful vested interests which will make it more and more difficult for the State in the time to come to do its duty by the tax-payer, while at the same time being fair to the manufacturer. It is a very serious thing to contemplate the immense monopoly which the right honourable Gentleman [Cunliffe-Lister] proposes to establish and strengthen in this country. . . . I think it right that Parliament at this time, when it is embarking on this prodigious expenditure, should have recalled to it the terrible misfortune that arose when we pursued this system in time of war.[17]

The question of compulsion raised by Dr. Addison was not a welcome one to the Government. Cooperation and goodwill were to be the touchstones of their rearmament policy, and any manner of compulsion of industry, they believed, would have a depressing effect on them. Cunliffe-Lister replied with some heat to Dr. Addison's assertions that excess profits would be made if powers to compel industry were not taken. He stated at one point: "Do not let us be so mealy-mouthed. I think the difference between the right honourable Gentleman and myself is that he has a passion for coercion, and I prefer to get effective action by agreement and good will." At another point he asserted:

I infinitely prefer to proceed by sound business agreements rather than by coercion unless coercion is proved necessary. . . . The question of powers can only be judged by results, and, as I told the right honourable Gentleman, I had every belief that I was going to be met by the industry in a perfectly fair manner, and I speak with the concurrence of those gentlemen, including Sir Hardman Lever, who are now advising me; and I do rather resent the suggestion that these firms will not play the game unless they are compelled.[18]

The only tangible result of this exchange was that the Government quietly began to insist that it have the right to inspect books where they pertained to work under government contract. Although this procedure was of no use in detecting overpricing by subcontractors or in computing the overall rate of profit of a firm, it was an

[17] *Parliamentary Debates* (House of Commons), Vol. 304, 7/22/35, col. 1615–1619.

[18] T 161/856/49179/1, 7/18/35, Extracts of parliamentary questions and answers.

indication that the Government was aware of the need to improve its oversight of the profitability of its contracts.

When the Air Force estimates were presented in the course of the 1936 budget debates, the issue of profiteering was again brought up. F. W. Pethick-Lawrence, speaking for the Labour Party, referred the House to the rather dramatic jumps in the prices of armaments-industry shares since 1935, and concluded:

> True, there has been talk of preventing undue profits, but there are no suggestions of any adequate steps to implement those promises, and it is clear that neither the City nor the armament firms have any belief that this limitation of profits is going to take place. . . . I think armament manufacturers have looked at the Government's promises of increased expenditure on defence, they have estimated the kind of limitations which the Government are likely to impose—they know their Tory Government just as we do on this side of the House, though from a rather different point of view—and they have come to the conclusion that very considerable profits are going to be the result of the whole of this policy which is adumbrated at the present time. I do not think they will be disappointed.[19]

While the Government did not deign to dignify these allegations with a reply in Parliament, *The Times*, in an earlier editorial on profiteering, had dismissed all such Labour charges, stating that

> there can be no excuse for [Labour] trying to distract public opinion from the urgent necessity of repairing our defences by insinuating that the programme is only a dodge to enrich capitalists.[20]

Although Pethick-Lawrence's observations were to the point, they caused no public outcry because there were as yet no facts with which to support them.

THE SHADOW SCHEME

While Baldwin and Swinton were giving repeated public assurances of the Government's resolve to prevent any excessive profit

---

[19] *Parliamentary Debates* (House of Commons), Vol. 311, 4/22/36, col. 162–164.

[20] Editorial, *The Times*, 2/27/36, p. 13.

making from the rearmament program, the Air Ministry had begun negotiations on two sets of contracts that were to determine whether or not that resolve would be effective. In early 1936 negotiations were opened with leading firms in the automobile industry with the intention of establishing the terms under which the automobile makers would agree to set up and operate shadow factories for the production of aircraft. This was the first government effort to implement the "Shadow Scheme," which Weir had recommended three years earlier, and he had a role in carrying it out. The other, more important negotiations were with the Society of British Aircraft Constructors, the S.B.A.C., on an industry-wide contract to cover the general terms under which all future government aircraft contracts would be executed. These negotiations were the acid test of the doctrine of cooperation, and put the lie to the professions of goodwill made by both government and industry. It is in the cold terms of contracts that future profits and losses are written, and in their negotiation quainter concepts, like the national interest, are set aside.

Lord Weir's plans to create additional war capacity by setting up shadow factories were predicated on the close collaboration of industry with government. In the D.R.C.'s Ideal Scheme the Government's rationale and approach were explained.

> We are convinced that it will only be possible to create the "shadow armament industry" in consultation with industry itself. Selected representatives of the industry and the organisations concerned in the expansion of industry for war production must be taken into confidence and the defence requirements problem explained to them by a Minister or Ministers, preferably the Prime Minister. It will be necessary to mobilise not only the good will of the manufacturers, but also their ideas and productive experience.[21]

In April of 1936 the Air Ministry called together the leading auto manufacturers, explained the urgency of the defence situation, and set forth its proposals for the creation of shadow factories to produce Bristol aircraft engines. Under the proposals the Government would provide the capital for the building of the new

[21] CAB 24/259, CP (36), 2/12/36, D.R.C. 37.

plant and would own it outright upon completion. The industry would supervise the building of the plant, which would be erected close to existing automobile factories, manage it, and hire and train the labor to operate it. For this service the manufacturers would be paid a fee based on output and efficiency. After the proposal was set forth there was much discussion of the organizational arrangement of the scheme, under which each shadow factory was to produce certain parts to be sent to a central factory for assembling. Certain of the owners wanted to build and assemble the engines from scratch in their own plants. Lord Nuffield, the owner and guiding genius behind Morris Motors, proved especially difficult in this regard. He eventually declined to enter the scheme as the result of some rather cavalier treatment by Swinton and Weir, and created a considerable public controversy over the management of the Air Ministry in the process.[22] Other firms, such as Rootes, had doubts about participating as well, dropping out only to reconsider within a week. The rather erratic resolve of some of the industrialists prompted a member of the air staff to write Swinton: "What an infernal nuisance these Captains of Industry really are! They are temperamental almost to the extent of being female."[23]

By late May the Air Ministry had a commitment from seven firms to join in the program, and the final negotiations were begun on the fee that was to be paid each firm for its management of its shadow factory. Swinton and Weir began their negotiations with Sir Herbert Austin, the founder and head of Austin Motors, the largest of the automobile manufacturers, on a contract for the shadow production of aircraft engines, intending the settlement to serve as a model for the contracts to be made with the rest of the firms. Things did not go smoothly. The Government began by offering to pay a yearly management fee of £17,500, a fixed payment of £150 for each engine produced, and a bonus of 10–15 percent of whatever savings were made in costs as production became more efficient. Leaving the bonus aside, this arrangement would mean Austin's making £152,500 if it produced 900 engines

---

[22] For a more detailed account of the Nuffield controversy see Reader, *Architect of Air Power*, pp. 252-269.

[23] Air 19/1, file on the creation of the Shadow Scheme.

in a year. Austin suggested that a more equitable arrangement would be for the government to pay a fee of £75,000, 10 percent of the cost of manufacturing each engine (which, given 900 engines at £400 each, would yield £360,000), and a 25 percent bonus on savings. This proposal, the savings bonus aside, would cost the Government £435,000, and left the two parties £282,500 apart. The give and take went on until a stalemate was reached with £90,000 still separating the two. At this point Swinton and Weir went to the D.P.R. and explained the situation, detailing the bargaining up to then. Swinton stated that: "unless settlement is reached on terms which the Government can approve, we shall be faced with the alternative of accepting terms which we think excessive, as the price of getting machines, which are an essential part of our programme, or attempting some form of control."[24] He then went on to point out the risks Austin would be taking in joining the scheme, and asked permission to go as high as £250,000 to close the deal. Although this was £65,000 more than the Government's last offer, Weir noted that it meant Austin would only be getting 5–7 percent of the £4–5 million worth of work he would be overseeing. From a businessman's point of view that was not very much. Chamberlain agreed to go along with Swinton, and the settlement was reached with Austin.[25] The agreements with the other firms were each negotiated separately, using the Austin agreement as a model, but their terms were not as generous, as it was argued that none of the other firms was as efficient as Austin. It is interesting to note that the auto manufacturers did not choose to negotiate through their trade association, the Society of British Motor Manufacturers, but instead elected to come to terms with the government individually. It was a decision they apparently regretted when they learned of the terms the Society of British Aircraft Constructors, S.B.A.C., had negotiated for its members.

THE McLINTOCK NEGOTIATIONS

To conduct the negotiations with the Air Ministry the S.B.A.C. hired Sir William McLintock, an eminent City accountant, and

---

[24] CAB 16/140, D.P.R. 86, 5/18/36, the finance of the shadow factories.
[25] CAB 16/136, D.P.R., 21st meeting, 5/25/36.

the contract agreement that eventually resulted from the negotiations was called the McLintock Agreement. The agreement lent form to a relationship between the industry and the Air Ministry that had been characterized by fluidity. At the end of the First World War there were a large number of aircraft firms, most of them small, that had been brought into being by the great demand for aircraft during the war. Civil demand for aircraft after the war was virtually nil, and the Air Ministry's needs in peacetime were also limited. As a result most of the firms folded, leaving only those to whom the Air Ministry chose to give contracts. In the early 1930s a "ring" of four firms produced aircraft engines and about fifteen firms produced airframes. All of these companies owed their survival to the Air Ministry, which in effect subsidized them by spreading its orders among them, insuring each enough business to survive. This was more than altruism on the Air Ministry's part, for it had an interest in seeing that the nation maintained the industrial capacity to produce aircraft. Although the technological capacity of the industry was quite advanced, as demonstrated by Britain's successes in the Schneider Trophy air races, its productive capacity, enervated by years of slack demand, left much to be desired.[26] In 1935, to put a newly designed model into production took five years, and for a large new plane, closer to eight. The conservatism bred by the years of struggling to survive did not dispose the industry to agree readily to sink capital into the expansion of plant and the introduction of new production techniques when the Government suddenly decided that the rapid expansion of the Air Force was vital to the national security. The aircraft firms approached the sudden new prosperity warily, concerned that the boom would be short lived, and that they would wind up, as so many did after the last war, with nothing to show for it but capital tied up in redundant plant. They had banded together to survive the hard times, and they stayed together to take the greatest possible advantage of the newfound prosperity. What the aircraft firms owed the Government for their survival was forgotten as they set out to insure that hard times would never come again. Security was their first concern, and until the Air Ministry guaranteed that the new demand for aircraft

---

[26] Fearon, "The British Airframe Industry and the State, 1918–1935."

would be long term, they would not move a finger to expand their plant. They refused to cooperate with the shadow scheme until it was promised that the shadow plants would operate only as long as the demand for planes exceeded the industry's capacity to produce them, and would be closed as soon as their capacity caught up with demand or vice versa. It was a seller's market, and they were resolved to make the best of it.

The members of the S.B.A.C. expressed extreme dissatisfaction with their treatment by the Government in a letter sent to the Air Ministry on March 18, 1936. They stated in the letter that, "the industry is so much perturbed by the present position as regards contract conditions, arrangements for fixing prices and other matters that a full and frank statement of the position is necessary." After elaborating on the efforts made by the industry to speed production over the past eight months after the Government suddenly decided that rapid expansion of the R.A.F. was imperative, the letter went on to say that, "it was assumed that there would be some reciprocity and that all steps would be taken to agree prices and deal with capital expenditure and other matters but in fact little, if any, progress has been made."[27] They blamed the state of affairs on poor Government organization and planning, and were insistent that the contract situation be straightened out before they committed themselves further to the expansion program. What they were insisting on were ironclad guarantees that plants in which they invested capital would not suddenly be rendered superfluous by a cutoff of the expansion program, and a clear statement of the rate of profit they could expect on their contracts. This letter with its implicit threats of non-cooperation set in motion the negotiations between the Air Ministry and the industry for an industry-wide contract.

A JUST RATE OF PROFIT

The industry's demands for a clear statement of the profit rate on Government contracts stemmed from the growing concern within the Government about what that rate should be and how it should be computed. In earlier times, before the increase in demand brought by the expansion program, "the basis [for the rate of profit] usually adopted on Air Ministry production contracts for air-

---

[27] Air 19/9, 3/18/36, letter from the S.B.A.C. to the Air Ministry.

craft and engines has been 10%. In view of the large increase in the turnover this rate may prove too high at any rate in some cases and the possibility of reducing the rate will need to be considered." It was this reconsideration of the basic profit rate, and the decision that it should be determined "in relation to increases of turnover involved after taking into account capital expenditure incurred on extensions,"[28] that stimulated the industry's concern in late 1935 and early 1936. In the face of the Opposition's persistent questioning of the Government's resolve to insure that profiteering did not occur as a result of the expansion program, and the Government's repeated assurances that it was taking every possible measure within the limits of the doctrine of cooperation with industry to see that such transgressions did not eventuate, the Government's sudden reluctance to give the aircraft industry a clearcut figure for the rate of profit on contracts is understandable. In late February Ashley Cooper, a member of the accounting firm of Mann, Judd, Gordon and Company, which the Air Ministry had hired to help write contracts that would prevent excessive profits, wrote Swinton:

> I read with grave misgiving the replies which the Prime Minister made in the House on the 20th and 24th on the control of profits on aircraft. I at once wrote to Judd, telling him that from my knowledge of the situation I was far from satisfied that undue profits were not being made. . . . I am convinced that you have not, as yet, a set-up at the Air Ministry to prevent undue profit making on the present programme, and you can therefore judge for yourself about the one to follow![29]

Squeezed by the Germans' rapidly increasing output of aircraft, the S.B.A.C.'s reluctance to expand production without a written contract delineating the terms, and the public's concern that under no circumstances should anyone be allowed to make excessive profits from the nation's need, the Government was forced to turn its attention to the question of determining an allowable rate of profit for the aircraft industry. On this subject there was a sharp split between the Treasury and the Air Ministry over whether the

[28] Weir Papers 19/16, 6/4/35, note by the Director of Contracts at the Air Ministry on prices and profits on Air Ministry work.
[29] Weir Papers 19/16, 2/26/36, copy of a letter from Ashley Cooper to Swinton.

rate of profit should be figured as a yearly percentage of the capital invested in a firm, or as a simple percentage of the cost of each contract. The ramifications of this difference in approach are significant. Under the Treasury's criterion the total profits earned during the year would be divided by the total capital invested to yield the rate of profit on the invested capital, and that would be the guideline for the rate of profit earned by the firm. Under the criterion demanded by the Air Ministry and the aircraft industry the rate of profit of a firm would be calculated by dividing the total profits earned during the year by the total value of the contracts it completed in that year, that is, its turnover. Although this difference appears to be simply one of accounting techniques, it lies at the center of the dispute over what did and did not constitute profiteering. An example reveals the significance of the difference between the two approaches: If a firm has a total of £2 million in private capital invested in it, and realizes £.5 million in profits on £10 million worth of contracts in one year, the rate of profit for that year, when computed under the Treasury criterion as a percentage of private capital invested, comes to 25 percent; whereas, when the rate of profit is computed as a percentage of the value of the contracts completed, it comes to only 5 percent. The second is clearly a justifiably modest rate of profit, while the first can just as clearly be seen as profiteering. The battles that focused on this issue both within the Government and between the Government and industry had a profound effect on the conduct of the whole rearmament program.

The Treasury position on the subject was clearly stated on March 18 in a note that turned out to be prophetic:

> We wish to put on record our view that the method of fixing profits in relation to production costs will almost certainly result in some firms making excessive profits, i.e. in what the public will justly regard as profiteering.[30]

At a meeting of Arthur Robinson's Treasury Sub-Committee on Contract Procedures it was decided that the Government should not become bound to any rate of profit it regarded as optimal, that the 10 percent that had been operative on Air Ministry contracts

[30] Weir Papers 19/16, 3/18/36, note by the Treasury Sub-Committee on Contract Procedure.

should be forgotten, and that contracts should be written as tightly as possible without reference to any given profit rate. The Treasury's concern in this regard was clearly stated by Neville Chamberlain, who stated at a meeting of Treasury and Defence ministers:

> If in fact at a later stage it was found that the dividends of munitions-making firms went up from 5% to 10% or even 20%, and there were stories in addition of hidden reserves, there would be a general loss of confidence, and the rearmament programme would be jeopardised.[31]

The Treasury was not particularly concerned with the morality, or even the cost in monetary terms, of profiteering, but was extremely sensitive to the political repercussions of its public revelation. Warren Fisher, especially, was concerned that such a revelation would result in the public's turning strongly against the rearmament program.

The concerns of the Air Ministry being different, caused it to view the question of the determination of profits in a different light. The Air Ministry's primary concern was with seeing to it that the nation had sufficient industrial capacity to produce the number of aircraft necessary for the national defence. It was the Air Ministry that had to deal directly with the S.B.A.C. and that had the most to lose by incurring the ill will of its members by insisting on terms that were unacceptable to the industry. The failure to procure the aircraft necessary for national defence would be laid at the door of the Air Ministry, regardless of the underlying problems, and for this reason, it was inclined to be more indulgent toward the needs and requests of the industry. To the Air Ministry the possible repercussions of the determination of the rate of profit in terms of turnover were far less significant than the realities of the lag between Britain's capacity to produce military aircraft and Germany's. Moreover the importance of retaining the cooperation and good will of the industry in the expansion program was deemed far greater than that of shaving a percentage point or two off the industry's profit rate. Where the Air Ministry was concerned with production and strategy, the Treasury was concerned with economics and politics.

[31] Weir Papers 19/16, 3/26/36, meeting of Treasury and Defence ministers on contracts.

The differences between the two ministries were brought into sharp relief by the negotiations between the Air Ministry and the S.B.A.C. In a letter to Chamberlain in early April Swinton is at pains to stress how difficult the negotiations with McLintock have been, and how stringent the terms of the Air Ministry are, especially on the question of profits. He then goes on to say that, "we all feel that we have gone as far as in fairness we can go, and an attempt to go further or change our principles would without doubt be resisted all along the line. . . ." The reference to changing principles alludes to the problem of the method of determining the rate of profit. According to the Air Ministry, and their advisers, Weir and Lever:

(1) It is impossible to segregate capital when a firm is doing business with the Government, work on civil contracts at home and on foreign contracts.

(2) If we attempt to ignore turnover, we abolish all incentive to production and efficiency. We are therefore strongly opposed to disregarding turnover, both in the interests of securing the programme and of reducing aggregate cost to the taxpayer.[32]

The Treasury view was that the Air Ministry was too sensitive to the problems of the aircraft industry, and not sensitive enough to the need to save the taxpayer's money. Moreover, the Treasury was unhappy with the failure of the Air Ministry to consult closely with it on the terms of the contract being negotiated with the S.B.A.C. The focus of the Treasury's concern, however, was the Air Ministry's position on profits. The nature of this concern is clearly conveyed by Chamberlain. At the bottom of a note by Weir on the question he wrote:

. . . I confess that all through the note I see the business man feeling the advantages of competitive tender and the naturalness of profits calculated on turnover emphasised to a degree which, whether it is right or wrong, does in fact envisage possibilities of profits which will not square with Government declarations.[33]

[32] Weir Papers 19/16, 4/8/36, letter by Swinton to Chamberlain on the profit clauses of the McLintock Agreement.
[33] T 161/986/40473/01/1, 3/27/36, Chamberlain comment on Weir note.

Despite the Treasury's persistent injection of its view on such topics as the determination of profit, the Air Ministry conducted its negotiations with the S.B.A.C. very much on its own. Under Government contracting procedure:

> The Treasury assumes general responsibility for laying down the principles to be observed in the placing of contracts, for regulating procedure and for dealing with any unusual conditions; while the contracting Departments take complete responsibility for contract administration in detail and for ensuring that the conditions attaching to particular contracts are appropriate and that the financial provisions are prudent and economical.[34]

While under these guidelines the Treasury could have insisted on a closer oversight of the bargaining with McLintock, the S.B.A.C.'s negotiator, Swinton and Weir felt that such an intrusion would limit their bargaining position and slow down the negotiations to the point that whatever goodwill with the industry remained would be lost. Even without the intrusion of the Treasury into the negotiations, the Air Ministry found that the S.B.A.C. was very difficult to bargain with. The aircraft firms knew what they wanted and were in a position to hold out until they got it. At one point the S.B.A.C. was on the verge of walking out of the negotiations and writing a public letter denouncing the Government's meanness and inflexibility. Weir, in notes for a letter intended to show them the imprudence of such action, set forth in his own language the enlightened industrialist's argument for cooperating with the Government in the rearmament program. Under the heading *"What do they want?"* he wrote in his own hand:

> If they once send a stupid letter a grave situation will have arisen. We must tell McL[intock] this + ask him to put before them [the S.B.A.C.] this—All we have proposed is fair + all our discussions have been in a fair spirit. No jewing. Weir, Lever, McL[intock] would not agree to anything unfair. We do agree it is not simple. Room for friction but that is the nature of

[34] T 161/942/44425, 6/5/36, Treasury Inter-Services Committee minutes on contract procedure.

the case. A stupid letter with threats would at once provoke [a] grave polit[ical]-indust[rial] situation + open up the whole case of private munitions supply. To blackmail the State in emergency is a present to Socialism, Communism + all evil movements. More than that, no sector of Pr[ivate] Enterprise can do this by itself, risking all Pr[ivate] Enterprise.[35]

This was, in the plainest terms, the argument for industry's not taking advantage of the existing Government in the rearmament program. The alternative to cooperation was compulsion, and compulsion meant either socialism or fascism, neither of which was attractive to the free market capitalist. Still the line between sabotaging the private enterprise system, and insuring that one received one's fair share was a narrow one, and where it was drawn depended very much on where one stood.

THE STRUGGLE FOR AN AGREEMENT

As Weir noted, the formulation of a general contract to cover the future orders for all types of aircraft production, including models that were not even yet on the drawing board, left room for friction. The central problem the negotiators had to deal with was how to settle on the price of planes that had never before been produced. Another issue of concern was the terms under which the Government could break off production under a contract before that contract had been completed. The necessity for stopping production could arise if a model under production became obsolete or too expensive. The clause that covered this contingency was called the break clause. There were also the questions concerning the Government's right to review the profits made on their contracts, the tax breaks to be extended on depreciation, the compensation to be paid by the Government in the event firms should become burdened with excess capacity, and many others. In the abstract such issues may seem tiresome and trivial, but because they were money issues involving contracts worth millions of pounds they were of the greatest importance to those involved. They are of importance to the historian as well, for they embody many of the factors that determined the direction taken by British rearmament.

[35] Weir Papers 19/24, undated notes.

On September 4, 1936, the Air Ministry sent the D.P.R. a resume of the tentative agreement that had been reached with Sir William McLintock. There were some remarkable compromises made in that agreement. For example, the Government's insistence on its right to review the profits made on its contracts and its demand that excessive profits be returned were balanced against the industry's insistence that it be compensated for any redundant capacity after the expansion program had ended. Under this compromise if a firm were found to have made excessive profits, and had claims against the Government for redundant capacity, it would be given the choice of either keeping the excess profits and not receiving compensation, or returning the profits and receiving the compensation. Of course the divisive issue of what constituted an excessive profit was not dealt with.

The central problem, as has been noted, was that of how the price of new models was to be determined. The method arrived at stipulated that the first two production runs, or batches, would be produced, and the actual production cost ascertained, on which the contractor would be paid an agreed, fixed rate of profit. On the basis of the experience gained from those two batches it was hoped that the contractor and the Air Ministry could agree on a fixed price to be paid for the product; if they could not, they would try to agree on a maximum price. If the production costs plus the agreed rate of profit on subsequent production runs were found to be less than the maximum price, the producer and the Government would split the savings. In the event that the cost plus the rate of profit exceeded the maximum price, the Government would pay the maximum and the producer would take the loss. Should the producer and the Air Ministry fail to reach an agreement on either a fixed or a maximum price, the firm would have the choice of either submitting the question to arbitration or going permanently on a cost plus agreed rate of profit basis, referred to then as a "time and line" basis.

When the Treasury learned of the terms of this arrangement, it was extremely displeased. It felt that Swinton had allowed himself to be maneuvered into a position in which the Government would have insufficient control over pricing. At the heart of the Treasury's objection was the producer's option to go on a time and line basis, referred to at the Treasury as "time and lime." This pricing technique had been the source of widespread abuse during the First World War, and was regarded with abhorrence within government circles; in fact, it had become almost synonymous in the public mind with profiteering. Its basic flaw was that it actually gave the producer an incentive to inflate his costs, because the greater the costs, the larger the amount of profit the percentage of those costs would yield. In the Treasury view the "costing clause" agreed to by Swinton

> in effect gives all firms concerned engaged in the production of aircraft and aero engines *a contractual right to refuse arbitration on contract prices, on two alternative bases, and to go on time and lime for the whole of their contracts*.[36]

Although the Air Ministry replied that it would grant a 10 percent rate of profit on agreed prices, and only 7½ percent on time and line, the Treasury felt that such an agreement left the Government vulnerable both financially and politically. If it became known that there were provisions for government contractors to produce on a time and line basis, the Opposition would make things very difficult both in the Commons and in the country. The Treasury concluded that the "time and lime" provision had to be renegotiated even though the S.B.A.C. was under the impression that the Government had agreed to that provision; in fact, of course, only the Air Ministry had agreed. The Government would give its assent only after the Treasury had expressed its satisfaction with the terms in question, and in this case the Treasury was anything but satisfied. The Treasury offered the Air Ministry two alternative approaches to reopening the subject with McLintock: the first was to balk when the general terms in which the existing agreement was expressed were being written into the agreement's final, specific form; the second was to confront the industry with the unpleasant fact that the contract as written would not enhance its already tarnished public image with regard to profiteering on arms. In an interesting marginal note on this Treasury memorandum Chamberlain wrote:

[36] T 161/1033/41460/1, 9/25/36, Bridges memorandum on the Air Ministry agreements with McLintock.

I'm afraid this rather serious departure [the time and line provision] arises from the S/S's insistence on being "his own Managing Director," a situation which creates a good deal of embarrassment for others.[37]

THE McLINTOCK AGREEMENT

Swinton did reopen the issue with McLintock, and they quickly agreed to a formula that obscured the time and line provision without really altering its substance. By mid-January the Air Ministry believed that it had arrived at terms that would satisfy the Treasury, and the final ratification of the McLintock Agreement seemed imminent. The Treasury, however, brought in its solicitor, who found legal problems with the agreement, and the negotiations dragged on through 1937. By late October the situation had reached an impasse. To A. H. Self, the Air Ministry's Second Secretary, it was

> becoming increasingly clear that the Industry desire to drift indefinitely so that they may secure in effect time and line contracts. This is sufficiently alarming from the Parliamentary aspect, but it is still more so in regard to available funds. Clearly the inefficient firms will benefit most under such an arrangement since they will be assured of a clear profit whilst getting full payment for their inefficient methods—moreover this is involving a set-back in production and delays in deliveries, which make aircraft almost obsolete by the time they are delivered. In short this method spells ruinous expense for inefficient equipment.[38]

When this note was brought to Swinton's attention, he gave orders to force the matter to a conclusion as soon as possible.

The industry's primary sticking point in this later bargaining was the Air Ministry's stipulation that the basic rate of profit on turnover should be 7½ percent rather than the usual 10 percent due "to the elimination of commercial risk when price is based on

[37] T 161/1033/41460/1, 9/25/36, Chamberlain's marginal note on Bridges memorandum.

[38] AVIA 10/241, undated but about 10/26/37, A. H. Self on the setting of aircraft prices.

actual costs." "We have had," wrote the Director of Contracts, Donald Banks, "the utmost difficulty in getting contractors to recognize this simple fact. . . ."[39] As late as April 1938 several firms were still balking at signing the McLintock Agreement prompting Self, Swinton, and Weir to warn that "it appeared that the firms, if they persisted in their present attitude, were heading straight for controls."[40]

The agreement was not finally signed until May 31, 1938, two years after the negotiations had begun. The contracts that were drawn up between aircraft firms and the Air Ministry in the intervening period basically conformed to the terms that had been agreed to by McLintock and the ministry at the time they were signed. Although the Treasury succeeded in forcing the Air Ministry to make substantive changes in the agreement that narrowed the industry's opportunities to earn excessive profits, the Air Ministry and the industry prevailed on the vital issue of the method by which the rate of profit was to be determined. Despite growing indications that aircraft firms were realizing exorbitant returns on their invested capital, profit continued to be calculated on the basis of turnover.

The prolonged and acrimonious nature of these negotiations did little to enhance the relationship between the Air Ministry and the S.B.A.C., which blamed all the holdups on the Air Ministry, not realizing that the Treasury was behind them; it also widened the split between the Air Ministry, which was trying to implement the program of industrial expansion for defence purposes, and the Treasury, which, it should be recalled, was the original champion of the cause of an expanded R.A.F. For its part the Treasury was quite correct in its objections. Although engendering considerable ill will, the Treasury was right in refusing to authorize a contract that would have permitted excessive profits. In this three way contest each party acted in accordance with its interests. The industry was naturally concerned with extracting the maximum profit and security from its favorable bargaining position. The Air Ministry was concerned with arriving at an agreement in the shortest possible time and with the least possible friction; concern with produc-

[39] AVIA 10/242, 1/15/38, Banks to Self on the profit rate on airframes.
[40] SWIN 270/4, 4/5/38, minutes of the 120th meeting of the Air Council.

tion disposed the Air Ministry to accede to the requests of the industry, with which it had to work, in the interest of fostering good will and cooperation. The Treasury, as the guardian of the exchequer, was naturally most concerned with the financial implications of the agreement, and made its objections on those grounds; it was also, not incidentally, the ministry most sensitive to the agreement's political ramifications. This is what the Treasury control is all about.

GOVERNMENT, LABOR, AND REARMAMENT

While the Government chose to pursue the doctrine of cooperation with business in its conduct of the rearmament program, its policy with respect to the other vital force in the economic community, labor, can be described as one of avoidance. This is not to say that questions concerning labor were not considered by the Government; they were, and in detail. Organized labor, however, was not consulted by the Government on any of the issues pertaining to it that came under Government consideration until the Government found its back to the wall in 1938.

Despite the fact that unemployment never dropped below 10 percent throughout the thirties, there was a shortage of skilled workers caused in part by the restrictive practices of the craft unions and in part by the inadaptability of a large segment of the unemployed to skilled work on account of age and remoteness from the centers where the skilled work was available. The introduction of the industrial expansion program in 1935 promised to worsen the situation. By February of 1936 the Minister of Labour, Ernest Brown, was writing that, "there are already indications that employers who cannot obtain labour elsewhere are beginning to raise their wages in competition with one another."[41] The Government wished to avoid the inflationary spiral that a persistence of such activity would produce. It felt that:

Wages in the armament industries must not be such as would attract labour from ordinary civil and export industries in such numbers as to impair their productive capacity; nor must wages in the industries be forced up in sympathy with wages in the

[41] CAB 24/260, CP (36), 2/21/36.

armaments industries, to a point that would destroy their competitive power in the world market.[42]

In addition the Government was concerned that such an increase in prices would make the rearmament program even more expensive than it already was.

Labor's concerns about the expansion program were very similar to those of industry. Members of skilled craft unions were worried that the armaments boom would be short-lived, and that if they relaxed their restrictions on access to their jobs, they would find themselves competing with the new workers for a dwindling number of positions when the boom ended. As Brown sagely noted:

> The trade unions for their part, will no doubt press upon the Government the argument that the men deserve as much consideration as the firms with whom arrangements are made for laying down of special plant and machinery, and that if the Government does not think it right that firms be left in the lurch and allowed to suffer loss of capital for having stepped into the breach, it should show similar consideration to the men in the matters beyond the unaided control of the employers.[43]

A trade unionist could not have put the case any better. The price of accommodating industry was already high, and the Government had no intention of giving the workers similar consideration regardless of the validity of their case.

The Government chose instead to concert with industry to try to regulate the demand for skilled labor in such a way as to preclude the kinds of sudden dislocation that result in large wage increases. The Government's labor adviser explained the situation this way:

> If a reasonable amount of time can be allowed, labour difficulties will be *eo tanto* reduced. If, however, it is decided that the maximum speed must be applied to the programme then a warning must be given that labour difficulties are probable. If the speed is such that the whole handling of the labour side of the situation can be left to the industries themselves, it will, from the point of view of the absence of labour troubles, be the

[42] CAB 24/261, CP (36), 3/26/36, Brown's paper on labor.
[43] Ibid.

best solution. The more the Government are directly involved, the more they will be put into the position of solving the employer's difficulties by buying off the Trade Unions.[44]

The Government was resolved to avoid involvement in such a situation, which could have not only financial but political repercussions. The Government's efforts to avoid dislocations of the skilled labor market involved the establishment of labor exchanges, which served the purpose of bringing workers and jobs together, and the careful placing of contracts to see that no area received more work than its workforce could carry out. The latter also involved the placement of contracts in such a manner that the contractor had sufficient time to hire and train his workforce, rather than buy already skilled workers away from another firm doing similar work. The object of this policy was to avoid the fostering of competition between Government contractors for labor, while at the same time avoiding entirely any dealing with the trade unions.

Throughout the discussions of labor questions there is a sense that organized labor is an adversary with whom the Government has no wish to become involved. There is more to it than the fact that the Labour Party that formed the Opposition had its base of support in the ranks of organized labor. Labor is spoken of and looked upon as "they," while the Government and industry, for all their disagreements, are "we." Nothing reveals this quite as clearly as this excerpt from a paper written for the Cabinet by Thomas Inskip.

> Direct action by the Government with the object, for example, of securing dilution [of labour] might well lead to very serious labour unrest and concessions to labour apt to damage the position of the export trade. I should anticipate that, long before the need for such action arises, ample warning will be available and there will be time to explore all avenues with industry and devise the least harmful solutions.[45]

Industry was consulted, cajoled, and treated with every consideration, while labor was regarded as and treated like a natural re-

[44] LAB 25/79, 3/9/36, Humbert Wolfe on labor and the Ideal Scheme.
[45] CAB 24/265, CP 297 (36), 10/30/36.

source that tended to be unfortunately expensive. The Government's suspicion that labor would exploit the needs of the nation in the interests of the working class meant that the assistance of a major productive force was denied the nation for two full years.

The Government could not have been unaware that, despite the parliamentary Labour Party's ongoing opposition to rearmament, as early as February 1936 the General Council of the Trades Union Congress had let it be known that it would be pleased to consult with the Government about the problems of expediting the rearmament process.[46] Although Bevin, Citrine, and Dalton struggled throughout 1936 and 1937 to alert the Labour Party to the importance of rearmament, and indicated the unions' willingness to cooperate in carrying it out, the Government chose to forego labor's assistance. It was not until the late spring of 1938, when driven by necessity, that the Government sought the cooperation of the trade unions; only then did it discover that its suspicions had been unfounded.

CONTROLS CONSIDERED AND REJECTED

In the course of 1936 the plans that had been made for the expansion of the productive capacity and output of the aircraft industry began to go awry. The shortage of skilled labor became increasingly acute, and the inability of the machine-tool industry to meet the demand for machinery for the new aircraft factories set production schedules back. In addition it became apparent that many aircraft firms had greatly overstated their productive capacity in order to attract Government contracts. When the situation came to the attention of Parliament in the late fall, it gave rise to renewed calls for the creation of a Ministry of Supply with powers to compel industry. By that time Government had already considered both the question of the compulsion of industry and that of a Ministry of Supply in some detail.

Although the Government had assured industry that the expansion program would not interfere with normal trade, at the January 13, 1936, meeting of the D.P.R. (D.R.) Baldwin, in response

[46] John F. Naylor's *Labour's International Policy, the Labour Party in the 1930's* (Boston: Houghton Mifflin, 1969), pp.152-155, 173-174 and *passim* provides a well informed discussion of the divisions within the Labour Party over the rearmament issue.

to an observation by Swinton that the Air Ministry program would require some "reinforcement" from civil industry, said that "it had always been assumed that industry must be interfered with to a certain extent."[47] By June it had become apparent that the levels of production and preparedness that were planned on under the existing programs would not be achieved within the designated time span unless those programs were accelerated. To be effective, acceleration would entail increased spending and quite possibly the control of industry. Weir, in a paper on the subject, noted the costs and benefits of control:

> Given powers of interference and control, the most valuable facilities—skilled labour and executive personnel—would become available to play an effective part in both production and organisation. On the other hand, interference in peace-time would produce entirely novel difficulties and dangers gravely affecting the financial and economic stability of the country.[48]

At the meeting of the D.P.R. (D.R.) where this paper was discussed, Weir pointed out that although controls would relieve the "bottle necks" that were holding up production, using them would be justified only if war were imminent and their use might prevent it. He then asked somewhat rhetorically, "If we interfered and took full powers to obtain acceleration of production, would not the economic consequences make war inevitable?"[49] In the context of the prevailing economic organization, his was a question of considerable validity, as the German example was to demonstrate. Weir did not feel that the existing situation merited either the creation of a Ministry of Supply or the use of controls. He suggested that better organization within the Government and the ordering of priorities in defence requirements would expedite the completion of the expansion program. Weir's close associate at the Air Ministry, Swinton, differed with him on this score. Swinton questioned how far it was possible to accelerate within the "existing machinery," and expressed the belief that should the programs continue to lag the creation of a Ministry of Supply with powers would be necessary. Duff Cooper was the only member of

[47] CAB 16/123, D.P.R. (D.R.), 1st meeting, 1/13/36.
[48] Weir Papers 19/2, 6/10/36, on the acceleration of the defence programs.
[49] CAB 16/136, D.P.R. (D.R.), 22nd meeting, 6/11/36.

the Cabinet to concur. Samuel Hoare expressed the view that "powers might come as a great shock to the country and result in an upheaval of industry."[50] Chamberlain pointed out that with or without a Ministry of Supply or controls, acceleration in itself would be expensive, possibly resulting in grave financial and economic consequences, and in his opinion the world situation did not merit either the risk or the expense.

By late October Parliament was again raising the question of the need for a Ministry of Supply as the result of the failures of the defence program. Much criticism was directed at Thomas Inskip, who, it will be remembered, had been appointed the Minister for the Co-ordination of Defence little more than eight months earlier in response to similar demands for a ministry of defence. It fell to Inskip to defend the Government's unwillingness to create such a ministry in the Commons. In a paper prepared for the Cabinet, in which he went over the arguments against the Ministry of Supply, he first reiterated the conclusion that had been reached in the Cabinet's earlier deliberations, that such a ministry without powers to compel industry would merely constitute another layer of bureaucracy through which defence plans would have to pass before being implemented. Such a reorganization would also cause additional delays in the existing programs. He then turned to the arguments for the Government's taking powers to control industry. He pointed out that even if controls were implemented to compel industries to turn over to defence work, it would be more than a year before those industries actually began production. If controls were implemented only over defence related firms and not over firms in the civil sector, it "would produce such grave discontent and indeed such a feeling of injustice that it could only be resorted to in situations of extreme necessity."[51] Such selective controls would constitute a form of taxation of the affected firms, which would have to forego some of their profits, while firms not connected with defence would be allowed to profit by the boom resulting from the increased defence spending. At the Cabinet meeting where Inskip's paper was discussed, Swinton gave a long, well-reasoned assessment of the possible uses of controls with or

[50] Ibid.
[51] CAB 24/265, CP 297 (36), 10/30/36.

without a Ministry of Supply. In speaking to Inskip's point that controls would not have any short-run effect on the rearmament program, Swinton pointed out that the arbitrary diversion of skilled labor to defence work would have an immediate effect on production because it would permit firms to use double shifts, where at the time many had only the manpower to operate a single shift. He described this approach as "picking the eyes out" of a factory. It meant that the Government could go to a firm involved in producing for the civil sector, where the skills of many of the workers were analogous to those required in defence work, and draft those workers for defence work. This action would result in a rapid and dramatic increase in defence production, but would also mean that the firms from which the skilled workers were lifted would probably shut down, leaving the unskilled workers in the factories unemployed. Such action on a wide scale would result in the rapid deflation of the consumer goods sector of the economy, and might well lead back into a period of economic depression. He concluded that such selective control, though effective in boosting defence production, would be unfair to both labor and business, and would be counterproductive. He then turned to a more plausible use of controls in which the Government, presumably under the aegis of a Ministry of Supply, would have the power to compel subcontractors to work for given contractors, and vice versa, as well as to compel subcontractors to produce specified goods. This plan would remedy the situation affecting some Government contractors whose subcontractors' products were in demand as the result of the expansion program. These subcontractors refused to accept orders to the specifications required by Government contractors, insisting that the contractors purchase the goods in the specifications in which the subcontractor was producing them or go elsewhere. In cases where the subcontractor was the only firm with the expertise to produce the good, this situation resulted in the loss of efficiency and time and created additional expense. Swinton concluded that such controls over selected contractors and subcontractors were justifiable because the firms affected would be those almost wholly involved in government work, whose profits came almost exclusively from defence contracts. The use of controls on this basis would also give those firms not under government power added incentive to cooperate voluntarily in defence work, thus precluding the need to extend controls any further.

The Cabinet's response to Swinton's proposals, which were probably trial balloons, was universally negative. The consensus was that any controls of the type suggested would disrupt the economy, dislocate industry, and were not warranted by the current situation. It should be made clear that Swinton was not advocating any of his proposals, but simply setting before the Cabinet the possibilities of increased defence production offered by controls. As the minister closest to the production problems the nation was encountering in its struggle to maintain air parity with Germany, he was most acutely aware of their intractability. He had apparently come to the conclusion that the degree to which the problems could be ameliorated within the limits dictated by the policy of non-interference with industry was insufficient to enable the country to keep pace with the Germans. He realized that control of industry was, at that time, a politically and economically unacceptable alternative, but believed that the time was rapidly approaching when the nation would have to take such steps or accept permanent aerial inferiority to Germany and the consequences inherent in such inferiority.

The problem that plagued Swinton, and converted him to the view that some form of control was necessary, was that Germany, through the use of all types of controls, was able to shift the nation's resources to the production of the goods it desired far more quickly than Britain could under the limitations imposed by her adherence to the precepts of the market economy. The crux of the problem lay in the fact that the invisible hand simply could not move resources as quickly as the dictator's hand. The technology of war had reached a point where the build-up of arms before the outbreak of hostilities was a major and possibly deciding factor in the outcome of those hostilities. Clearly a nation that could use controls to marshal her resources for the production of arms over a short period of time could create a military establishment that would enable her to dictate militarily to a nation that eschewed controls until such time as she was able to move her economy to the production of arms. Put simply, a command economy required much less time than a market economy to gear up for war. The nation with the market economy was thus vulnerable, and this was Britain's plight in the late thirties.[52]

The significance of the Government's decision to carry out the program of rearmament without resort to controls should not be understated. It affected the pace at which Britain rearmed just as surely as did the Treasury's insistence that orthodoxy be maintained in the financing of rearmament. The Government found the risks to national security entailed by that decision far more acceptable than the risks to the nation's economic and political stability entailed by the alternatives.

[52] Since the Second World War this problem has been resolved by the expedient of the world's dominant private enterprise economy's maintaining a full scale armaments industry. Thus, the kind of situation in which Britain found herself in the thirties has been obviated, but at serious economic, social, and political cost.

CHAPTER IV

The Financing of Defence

PRODUCTION PROBLEMS

B Y early 1937 the Government's rearmament program had been under way for almost a year. Thomas Inskip, in a review of the progress, asserted that it "is undoubtedly behindhand in many important respects, but it is behindhand just in those areas which no compulsory powers could affect, such as guns, tanks and technical instruments like predictors."[1] He went on to qualify his statement by saying that the shortage of skilled labor was at the heart of much of the problem, and that short of taking up Swinton's proposal to "pick the eyes out" of certain civil industries by taking powers to draft their skilled workers, nothing more could be done. The aircraft industry, which was under the greatest pressure in the expansion drive, was having the most trouble completing its contracts. Its optimistic delivery promises were severely set back by the shortages of skilled labor and machine tools, as well as by its own poor organization. There were few production problems for the Admiralty, and those of the Army were more attributable to the Government's indecision about the role the Army should play than to industrial problems. Britain had reached a point at which, in the words of Arthur Robinson, the Chairman of the Supply Board, "unless statutory interference in industry is adopted little more can be done to affect supply problems drastically. . . . We are really at peace, but we are rearming as fast as we can without producing irretrievable economic difficulties."[2] Taking note of the production problems at a Cabinet meeting, Chamberlain observed that they "showed that even the present programmes were placing a heavy strain on our resources. Any additional strain might put our present programmes in

[1] CAB 24/264, CP 40 (37), 2/1/37, Inskip on progress in meeting defence requirements.
[2] Weir Papers 18/5, 4/5/37, Arthur Robinson on the state of rearmament.

'When I have obtained, as I shall obtain, some sort of a statement which may be satisfactory as to basic principles, I hope that I shall be able to proceed with the practical questions.'— *MOMENTOUS MOUTHFUL FROM TOMMY INSKIP.*

3. "Big Arms Speed-Up"—Low's comment after one of Inskip's ineffectual replies in Parliament to a question about the progress of the Government's rearmament effort. Colonel Blimp assists Inskip. *Evening Standard*, March 29, 1937.

jeopardy."[3] Chamberlain and the Treasury were more concerned about the impact of the rearmament program on the economy and its cost than about the difficulties of making it effective. This statement was a hint of the growing resolve at the Treasury that the growth of rearmament had to be curtailed.

THE TREASURY VERSUS THE ARMY

The Army had long been the object of the Treasury's attentions when it came to pruning the defence budget, and in 1937 the battle between the War Office and the Treasury was even more heated than usual. In December 1936, when the preliminary estimates were submitted for the upcoming budget, Duff Cooper reopened the question of the role of the Army and the part the Territorial Army, the T. A., was to play in it. Alluding to Britain's weakness in the Rhineland crisis, he pointed out that at the present time the Army could send only two divisions abroad, and that most of their equipment was on the standard of the First World War. All the cavalry and transport was horsed rather than motorized, the Army had no tank force, and it lacked its quota of Bren and anti-tank guns.[4] Although the Army was making every effort to mechanize, budget restrictions were slowing them down. More important, owing to production lags and lack of capacity, the supply of Bren and anti-tank guns was going to be short for another year or more.[5] Duff Cooper's primary concern, however, was to persuade the Cabinet that the Territorials should be fully equipped. He contended that in any war situation on the Continent the Field Force would be completely engaged, and the T. A. would have to take its place in the garrisons around the Empire, as well as provide them with reserves against their losses in the field. If the Territorial Army were not equipped, it would be unable to fulfil those roles, and the Army would be ineffective as a consequence. Without the T. A., equipped to provide reserves for the Field Force, the Field Force itself could not be committed to fight-

[3] CAB 23/87, 5 (37) 11, 2/3/37.

[4] Basil Liddell Hart in his *Memoirs of a Captain* (London: Cassell, 1965), Vol. I, pp. 261-276, points out that in the 1935-36 Army budget £400,000 was to be spent on forage for horses and only £121,000 for gasoline.

[5] CAB 24/265, CP 325 (36), 12/4/36, Duff Cooper on the Army controversy.

ing, for it would be decimated within a matter of weeks.[6] He contended that because of the amount of time required to create the industrial capacity to produce arms, it was necessary to get authorization to place orders immediately if the T. A. were to be ready by 1939. His basic conflict with Chamberlain was over the question of whether the nation could manage such a program without industrial and economic complications. Duff Cooper felt it could, but the Treasury argued that shortages of labor and raw materials would become real problems.[7] When the issue came up for discussion in the Cabinet on the sixteenth of December, the arguments were familiar. Chamberlain suggested that the policy of sending the T. A. abroad be reconsidered, and that the nation's resources would be better used if the Air Force was expanded and authorized to be used in its stead. Inskip spoke in support of Duff Cooper's position, and the First Commissioner of Works, Earl Stanhope, pointed out that the experience in China and Abyssinia suggested that a war could not be won by air power alone, and that it should not be considered a deterrent in itself.[8] The Treasury had only reluctantly acceded to the Cabinet's decision of the previous February to equip the Army to fight on the Continent, and adamantly refused to accept the War Office's contention that divisions of the T. A. had to be fully equipped to serve as reserves if the Army were to be effective in its role. Until the dispute over the role of the T. A. was resolved, the Army estimates could not be settled and, more importantly, orders could not be placed for the equipment the Army so badly required. Army contractors were reluctant to accept the contracts the Army was authorized to offer for fear they would be cancelled when the Cabinet reached a decision on the Army's role.

In January the Treasury made a subtle but very effective move to force the War Office to give up their battle to have the T. A. equipped. They persuaded the C.I.D. to command the War Office to prepare a complete conspectus of all the Army's requirements, and relate them to the existing industrial capacity of the nation. Until this conspectus was completed the Army was not to be al-

[6] CAB 24/265, CP 326 (36), 12/3/36, Duff Cooper on the role of the Army.
[7] CAB 24/265, CP 334 (36), 12/11/36, Chamberlain on the role of the Army.
[8] CAB 23/81, 75 (36) 6, 12/16/36.

137

lowed to order any additional equipment including much needed anti-aircraft guns. When completed, the requirements set forth in the conspectus were to be listed in order of priority by a joint committee of the War and Foreign Offices, and finally submitted to the C.I.D. for consideration. The Treasury hoped to thus end the controversy over the equipping of the T. A., at least as far as the current budget was concerned, for it knew the Army would require at least several months to draw up the conspectus. There is every reason to believe that the Treasury was calculating as well that, by the time the Army had completed the task, Neville Chamberlain would be the new Prime Minister and the whole question of the Army's role would be reconsidered under circumstances more favorable to the Treasury's point of view.

The Army, however, refused to go along with the Treasury's ploy. Both Duff Cooper and the C.O.S. protested the C.I.D. order, pointing out that, until the Cabinet made a final decision on the Army's role and the T. A.'s part in it, the Army had no grounds on which to draw up a conspectus. Furthermore, they drew the Cabinet's attention to the fact that the C.I.D. order would set back the production of goods the Cabinet had recognized as vital, such as the anti-aircraft guns, at least six months. The Cabinet, unable to arrive at a decision on the matter, turned it over to Inskip, who was to draw up proposals and go over them with Chamberlain in an effort to reach a compromise agreement acceptable to the Treasury.

The choice of Inskip as arbiter in this dispute marks the beginning of the span of time during which he was to become an increasingly influential voice within the Government on questions concerning rearmament. Among the first to recognize this was Maurice Hankey, who saw his own influence declining as Chamberlain moved closer to assuming the leadership of the nation. In a note to Inskip, in which he expressed his strong belief that the Army should be fully equipped to meet its Continental responsibilities, he suggested that Inskip raise the points dealt with in his note under the guise of his own proposals in his discussions with Chamberlain, "as the Chancellor of the Exchequer probably regards me, as Chairman of the Defence Requirements Committee as one of the 'old gang' on this subject. Anyhow he has never asked my opinion on it, so I do not suppose he wants it."[9] As we shall see, Hankey and Inskip developed a working relationship that permitted Hankey to continue to influence the formulation of defence policy despite Chamberlain's lack of interest in his views. This relationship also benefited Inskip, providing him with independent, highly informed insights, which were of great value in the increasingly partisan struggle between the Treasury and the services he was called upon to arbitrate.

Inskip set forth his proposal on the role of the Army in a paper that was presented to the Cabinet on the second of February. In it he expressed his agreement with the view held by the Army and Hankey that the T. A. was an integral part of the continental force, and should be equipped in a manner that would allow it to act as such. As it was a reserve force, there was, to his mind, no need to equip it as fully as the Regular Army, nor to maintain it at the same level of preparedness. It was his "instinctive feeling that we simply cannot sustain a large Army in peace while we maintain a very powerful and modern Navy and an equally powerful and up-to-date Air Force. . . . If we cannot afford to do this, I see no object in treating this as if it was the ideal to be aimed at."[10] He proposed that enough equipment be provided to outfit one or two T. A. divisions, which in peacetime was to be spread among ten T. A. divisions for training purposes. In the event of war the equipment would be reassembled for the one or two divisions, while the others would wait until industry could gear up sufficiently to provide for them. The Treasury accepted the compromise as did the War Office, and the Cabinet duly authorized it. This, however, was not quite the end of the story. Late in April Duff Cooper presented another paper to the Cabinet asserting that at least part of the T. A. should be ready to support the regulars within four months. (Under the Inskip plan this support could not be provided in less than six months.) He requested that the equipment for four T. A. divisions be provided by April 1941. In

[9] CAB 63/52, MO (37) 1, 1/18/37, Hankey to Inskip. Although Hankey realized that Chamberlain's accession to the premiership would diminish his influence, he welcomed it because he believed that Chamberlain would provide the rearmament program with the direction it lacked under Baldwin. Roskill, *Hankey*, Vol. III, p. 264.

[10] CAB 24/268, CP 46 (37), 2/2/37, Inskip on the role of the Army.

the Cabinet meetings that ensued, Inskip supported Duff Cooper's demand, asking as a compromise for authorization for at least the equipment of two T. A. divisions on a full-time basis in order that they could support the regular Army in the requisite period. He went on to point out that the issue had been hanging fire since the previous December, and a decision had to be made if the equipment was to be forthcoming within a reasonable amount of time. Neither Chamberlain nor the Treasury was happy about the new proposal but they acquiesced. At the Cabinet meeting where the compromise was authorized, Chamberlain stated that he "definitely did challenge the policy of their [the Government's] military advisers."[11] He questioned once more the whole assumption that Britain must prepare to fight on the Continent as well as in the air and on the sea. He felt that the nation simply could not wage the next war in the manner in which she had waged the last one, and that it was up to Britain's Continental allies to contribute the armies necessary to their own defence. He added that the whole question was open to re-evaluation in the future, an indication of things to come when he would take office less than a month hence.

NEW AIR MINISTRY PROPOSALS

While the Army was struggling with the Treasury for funds, the R.A.F. presented plans that would significantly expand the program initiated in May 1935 and authorized the following February, which had come to be called Scheme F. In presenting the new plan, referred to as Scheme H, the Air Ministry stressed that, according to intelligence, although the R.A.F. was equal to Germany's Air Force in numbers at the present time, Germany's newly created capacity to produce aircraft would put Britain at a serious disadvantage by 1939 if she adhered to Scheme F. The Air Ministry projected that by April of 1939 the Germans would have 800 fighters and 1,700 bombers, while under existing plans Britain would have only 730 fighters and 1,022 bombers.[12] The Air

[11] CAB 23/89, 20 (37) 4, 5/5/37.
[12] CAB 24/267, CP 18 (37), 1/14/37, Air Ministry proposals for the expansion of the R.A.F.; CAB 23/87, 5 (37) 12, 2/3/37, secret intelligence supplement on German air strength submitted at the Cabinet meeting.

Ministry plan would enable Britain to just about match the German figures at an additional cost of £50 million over the upcoming four years. Another aspect of the Air Ministry's proposals was its suggestion that the Government abandon the numerical criteria for air parity with Germany upon which Baldwin's 1934 pledges to the nation had been based. The ministry suggested that more meaningful criteria were: a. the creation of a bomber force with a range and payload capacity equal to Germany's; b. a fighter force of sufficient size and strength to meet and repulse the anticipated scale of attack. The crude numerical criteria were both misleading and potentially embarrassing politically, the Air Ministry explained. They were misleading for two reasons. First, the Germans had built many planes designed solely to provide support for the army, which lacked the range to pose any threat to Britain. Second, the Germans required many more fighters to protect their far flung industrial centers than Britain did to protect her own comparatively closely bunched urban areas. Thus, the Air Ministry argued, Britain could have an air force that posed the same threat to Germany as Germany's posed to her without having anywhere near an equal number of aircraft.[13] It was suggested by the Air Ministry that the Government shift its public position away from the numerical basis as subtly as possible after Baldwin, who had given the numerical pledges, stepped down.

The Air Ministry's interest in seeing this change in criteria carried out was by no means based solely on concern about its possible political ramifications. The whole R.A.F. philosophy centered on the importance of the bomber, and implicit in the proposed criteria was the notion that the bomber force should be expanded while the number of fighters should remain fairly static. While the numerical criteria tended to put the emphasis on fighters—for one fighter counted as much as one bomber and was far easier and cheaper to build—the new criteria turned the emphasis around, and freed the Air Ministry to concentrate on building what it considered its most potent weapon. The bomber was to an air marshal what the battleship was to an admiral.

Scheme H reflected the Air Ministry's preoccupation in that it

[13] CAB 23/87, 4 (37) 3, 1/27/37; CAB 24/267, CP 27 (37), 1/22/37, Swinton review of Baldwin parity pledges.

projected the production of 600 more bombers and only 50 more fighters than were to be built under Scheme F. The Treasury was not anxious to incur the cost of this rather expensive program, although it had no quarrel with the reasoning behind it. It contended that the nation could not afford it, and on February 24 the Cabinet agreed.[14] All the Air Ministry gained was approval to recruit and train pilots and mechanics, and to purchase land in order to begin building new aerodromes. This marked the first rebuff of an Air Ministry request for funds, and it was justifiable. There is every indication that the Air Ministry made its proposals to take advantage of the concern about the threat from the air in order to get its pet programs into production. This maneuver did little to improve its relationship with the Treasury, a relationship that had already been weakened by what the Treasury regarded as the Air Ministry's profligacy in the negotiation of the contract agreements with McLintock.

Not to be left out, the Navy pressed once again for the New Standard fleet, stressing that failure to approve it was tantamount to writing off the Empire in the Far East. The Cabinet continued to approve the program in principle, while refusing, at the insistence of the Treasury, to fund it. Efforts were also initiated by the Government to assure New Zealand and Australia of Britain's continuing commitment to their defence, but at the same time to encourage them to undertake more naval building on their own. Meanwhile the Admiralty patiently waited for its opportunity to have the New Standard approved.

TREASURY REACTION TO THE DEFENCE ESTIMATES

Although the Treasury succeeded in making massive cuts in the service's preliminary estimates for the 1937 budget, the impact of the size of those preliminary estimates staggered the Treasury to the point that it began to reassess completely both the rearmament program and its financing. The figures that caused Treasury dismay were these. The cost of the 1936 defence budget had been £180 million, an increase of more than £60 million over the previous year. The projected cost of the 1937 budget, based on the

---

[14] CAB 23/87, 9 (37) 5, 2/24/37.

programs that had been authorized, was £211 million. However, the estimates actually submitted to the Treasury for the 1937 defence budget came to a stunning £286 million, £100 million more than the previous year. Edward Bridges wrote that "this is a good deal higher than anything which I anticipated in my gloomier moments,"[15] but held out hope that that figure could be reduced. His colleague Richard Hopkins was less optimistic. He estimated that the total budget would have an increase of £120 million over the previous year, and commented, "From this deduct Mr. Bridges' hopes of cutting down the Defence figure. To this add provision for Defence Supplementaries during the coming year as new ideas strike the General Staff. . . ."[16] At a Cabinet meeting Chamberlain explained the Treasury's anxiety, and gave some hints of the direction Treasury policy on defence issues might take as a consequence.

> The Chancellor added that he was getting concerned at the mounting cost of the [defence] programme. It was difficult for him to take a line in opposition in this question of national safety, but he wanted the Cabinet to realise that the cost was mounting at a giddy rate. The original estimate of £400,000,000 was already far exceeded and programmes were constantly increasing. Before long, he thought, people would be talking about an unbalanced Budget, and we might find that our credit was not so good as it was a few years ago. He said this because while recognising that national safety came first, our resources were not unlimited and we were putting burdens on future generations. . . . he wanted Government Departments not to think that because a heavy expenditure was being incurred, this was the time to slip in developments of convenience which had been refused in the past.[17]

This statement signalled the onset of the Treasury's campaign to bring the rearmament program under tighter rein. The Treasury felt that its efforts in the course of 1936 to relax some of the its control over contracting in order to speed the defence programs, in addition to the authorization of a number of new programs in

[15] T 171/332, 11/27/36, Bridges on the defence estimates.
[16] T 171/332, 12/7/36, Hopkins on the budget.
[17] CAB 23/86, 63 (36) 4, 11/4/36.

the course of the year, had fostered an atmosphere in which the services felt that there were few limits to what could be gotten in the name of defence. The Treasury sought to reassert its control over defence spending, and in so doing give rearmament the direction it felt was lacking. The Treasury drive began in the budget battles with the services and came to full flower with Neville Chamberlain's assumption of the premiership.

BORROWING FOR REARMAMENT

The immediate issue facing the Treasury was how to pay for a rearmament program that was going to become increasingly costly even if carried out with optimal efficiency—this was the first question the service estimates raised. It had been clear since late 1935 that the exceptional expenses of rearmament would have to be met by borrowing in some form, and as we have seen, at that time the Treasury set forth the criteria under which it felt borrowing was justified. Although they did not relish the prospect, everyone at the Treasury realized that 1937 had to be the year in which a borrowing program was instituted. In late November 1936 Frederick Phillips wrote a rather intriguing note explaining that if the Government did not wish to reveal that it was borrowing, there were ways to hide the fact in an apparently balanced budget. He wrote:

> The position is that we are budgeting for a deficit, that the pretence of producing a balanced budget is a fiction, and that there is no great technical difficulty in producing for a series of years budgets which are balanced at the end of the year to the nearest penny. I see no reason to doubt that this result or something close to it could be produced by the method proposed by Mr. Ismay [another Treasury economist] which involves merely an adjustment inside the Exchequer books at the end of the year. Perhaps a half a dozen financial writers in the country would understand from published accounts what was happening, but I doubt if any one of the half dozen is capable of making the position clear to the public.[18]

The Government had no desire or reason to go this route, how-

[18] T 160/688/14996/2, 11/27/37, Phillips on the budget and defence.

ever, and the suggestion was set aside as the Treasury got on with the task of preparing to float a defence loan.

The considerations that went into this task were several, the first being an estimate of how much would have to be borrowed and over what span of time that borrowing should be spread. Phillips foresaw defence spending reaching £320 million a year at the height of the rearmament program, and he felt that at least £120 million of the total would have to be raised by new taxes and borrowing. In the current year he could not foresee more than £180.5 million of the £280.5 million estimated by the services as necessary for defence coming out of revenue, leaving a £100 million deficit. The Treasury decided to increase taxation in keeping with the 1935 stricture that taxation would have to be raised if the borrowing were to be justifiable. In Hopkins' words, "the shortage is so great that, as it seems to me, it would not be decently possible to resort to borrowing first without a concurrent increase of taxation."[19] The Treasury then proposed to ask Parliament to authorize the Government to borrow £400 million over a five-year span at a rate of £80 million a year.

The Treasury was particularly uneasy about the reception of such a loan in the financial community, where its success or failure would be determined. As early as February 1936 rumors had been circulating in the city concerning the possibility of a defence loan being floated during the next year, and the financial community had reacted negatively to them. There was widespread concern that such a loan would be inflationary, and the banking community was particularly apprehensive that the loan would drive up interest rates.[20] Against this background Phillips wrote in January 1937:

> We have really no experience of borrowing *new* money since the war; even in bad years the worst that could be said against

[19] T 171/332, 1/18/37, Hopkins note written on a Phillips memorandum about the projected revenue for 1937.

[20] *The Economist* led in reporting the concerns of the City about a defence loan, and came out strongly against it editorially. It raised the possibility in an article entitled "Defence Loan Rumors," 2/15/36, p. 352, and ran two lead articles detailing the City's opposition to such financing, "The Boom and the Budget" (pp. 157-158) and "Bankers and the Boom" (pp. 221-222), on January 23, 1937, while

our credit was that we were not reducing the debt. I doubt very much, with the gilt edged market in a jumpy and rather dangerous mood, whether public opinion would be satisfied unless a substantial proportion, say £30 million a year, were raised by taxation. By public opinion I mean of course the opinion of the people who are going to lend us money.[21]

The most important of those who were going to lend the Government money was the Bank of England, a privately owned institution that had held its charter as the government's sole banker for nearly two and a half centuries. One of the Bank's functions in its role as the government's banker was to act as the residual buyer for undersubscribed issues of government loans. Because this function involved some financial risk to the Bank, it technically had the right to refuse to act as the residual buyer if it felt the terms of the loan posed too great a financial risk. Therefore the Treasury always consulted with the Bank and secured its approval of the terms of a proposed loan before giving the Cabinet any indication that a loan was under consideration, although in fact the Bank had never incurred any losses in its capacity as residual buyer up to this time.[22]

On February 4 Chamberlain went to the Bank to meet with the Governor, Montagu Norman, a man about whom much is rumored but little is known, despite the fact that he probably wielded more power over British finance in the interwar period than any other individual.[23] After listening to Chamberlain's de-

---

the Treasury was trying to decide how to make such a loan more acceptable to the City.

[21] T 161/783/48431/02/1, 1/19/37, Phillips note on finance and future defence spending.

[22] T 175/95, file. During the ensuing two years the Bank did take a loss on the operation and asked the Treasury to seek legislation to rectify the situation.

[23] There are two biographies of Norman, neither of which touches on the political considerations underlying Bank policy. Henry Clay's *Lord Norman* (London: Macmillan, 1957) is the official biography issued under the aegis of the Bank, while Andrew Boyle's *Montagu Norman: A Biography* (London: Cassell, 1967) was commissioned by Lady Norman to fill in the personal dimension she felt the Clay biography lacked. It was a dimension that Boyle had considerable difficulty locating. The basic problem with trying to assess either Norman's or the Bank's influence on Government policy is that, although the Bank is now a public institution, its archives for the period are not open to scholars, nor are most of the Treasury files pertaining to its business with the Bank.

scription of the proposed National Defence Loan, Norman expressed the belief that its magnitude would shock the financial community, and that it would have a depressing effect on government credit. The memorandum describing the meeting took pains to note that "he [Norman] remained quite calm on the subject, but raised two points." First, he questioned why the rate of interest (3 percent) on the loan was to be so low and expressed his doubts that it would be very well subscribed at that level—a matter of immediate concern to the Bank, which would have to support the bond issue if the rate were too low. Second, he "thought it most undesirable to state in advance the amount of money we proposed to borrow next year. This would have, he thought, most damaging effects upon the market which would go utterly stale."[24] Chamberlain explained that the Government was legally bound to disclose the size of the loan. Richard Hopkins apparently had subsequent meetings with Norman, which are alluded to in the Treasury papers, but of which there is no available record. Suffice it to say that, when the borrowing plan outlined to Norman was announced in Parliament on February 11, it shocked the securities market as he had said it would.[25]

THE NATIONAL DEFENCE CONTRIBUTION

Within a week of Chamberlain's meeting with Norman there appeared at the Treasury a note from E. R. Farber of Inland Revenue, the Treasury's tax division, dealing with a suggestion made to him by Chamberlain that he work out a tax based on the increase in profits made by firms as the result of the prosperity brought about by rearmament. This note marks the first mention of what was to be the most controversial fiscal measure to be introduced in Britain in the thirties, the National Defence Contribution, usually referred to as the N.D.C. The genesis of the N.D.C. is unclear. It is probable that it was conceived by Chamberlain to meet the needs of the Government for additional tax revenue to make the National Defence Loan appear more fiscally responsible, while at the same time meeting the anticipated Labour objec-

[24] T 172/1853, 2/5/36, a note from Chamberlain's personal secretary, J. H. Woods, to Hopkins and Phillips describing the meeting with Norman.

[25] The City's reaction to the proposal and the reasons for it were discussed in a lead article in *The Economist* entitled "War Loans Again," pp. 341-342, 2/13/37.

tions that the worker was being made to pay for rearmament in the form of tax increases while the industrialists were reaping unconscionable profits from it. The timing of the measure, however, leaves open the possibility that Norman might have made the institution of such a tax the condition on which he agreed to allow the Bank to take part in the issuance of the loan. His rationale would have been that something had to be done to make the National Defence Loan more attractive to the investor than investment in the industrial sector. A tax on profits might well have had that effect. From the evidence available one would have to conclude that the N.D.C. was Chamberlain's brainchild, although records of Hopkins' meetings with Norman might prove otherwise.[26]

The rationale behind the N.D.C. was that businesses that had profited from the prosperity that came in the wake of the rearmament program should bear some of the burden of paying for rearmament out of their increased profits. The measure's proponents at the Treasury believed that such a tax would demonstrate the nation's resolve to be responsible in its borrowing, thus maintaining international confidence in the pound and reassuring subscribers to the National Defence Loan of the nation's ability to repay it. The beauty of the tax from the political point of view suggests that Chamberlain or someone else in the Treasury conceived it. As Chamberlain pointed out to a group of important industrialists who had gone to the Treasury to voice their objections to the tax, if industry desired the current unprecedented period of industrial peace to continue, it was vitally important that industrialists should forestall any impression that they were not carrying their fair share of the burden of rearmament. If such an impression should develop, labor would begin to strike, everyone would be worse off, and the defence program would be held up as well. He added that, "We might end up with such concessions to labour as would very seriously handicap our competitive trades."[27] What

[26] In the course of an attack on the N.D.C. in Parliament, Robert Boothby, a Conservative MP, commented: "I cannot believe that these proposals were really devised by the right honourable Gentleman the Chancellor of the Exchequer himself; I still have an idea that the Bank of England had something to do with them." *Parliamentary Debates* (House of Commons), Vol. 323, col. 291.

[27] T 172/1856, 5/13/37, minutes of a meeting of a special committee of industrial organizations with Chamberlain about the N.D.C.

the Government feared most was that labor would begin to demand wage increases to match the increased profits business was enjoying, a situation that would result in an inflationary spiral and a decrease in productivity, both of which would have adverse effects on the stability of the pound. The actual amount of the burden of rearmament that industry was being asked to carry was very small, "only a ripple upon its [the defence budget's] surface,"[28] to borrow Chamberlain's metaphor. It was so small in fact as to make almost inescapable the conclusion that the motives behind it were political.

Chamberlain had carefully weighed these arguments while considering whether or not to propose the tax and had apparently concluded that the businessmen who were to bear it would consider the financial burden a small enough price to pay for continued labor peace. Although he was aware of the political risks he was taking, he believed that the force of his arguments would win over the critics. Because of the secrecy involved in the preparation of the budget, however, the Treasury was unable to consult with business before the budget was presented to gauge its reaction to the proposal. Consequently both Chamberlain and the Treasury were taken aback by the severity of the reaction against the N.D.C. from the Conservative backbenches and the business community after the tax was announced in the course of the budget speech on April 20, 1937.[29] *The Manchester Guardian*'s political correspondent described that reaction in the following terms.

> Astonished Tory industrialists considered the new duty had turned a hitherto safe and respectable Budget into a Socialist Budget, Socialism being to them of course not a political philosophy but pure brigandage. . . . The boldness of the step will appear later. The Tory industrialists are the most powerful section of the party, and people within the fold are declaring tonight that they must offer pained opposition to the duty.[30]

[28] *The Banker's Magazine*, Vol. 143 (June 1937), pp. 925-934, the text of Chamberlain's speech to the British Bankers Association given on 4/29/37.

[29] Drawn in a manner intended to tax most heavily those who profited most from the surge of prosperity that the Government believed was directly attributable to the increase in defence spending, the N.D.C. was to be calculated as a percentage of the increase in a firm's rate of profit over its rate of profit in a base period between 1932 and 1936.

[30] *The Manchester Guardian*, 4/21/37, p 11.

Especially surprising to the Government was the opposition of
the financial community to the measure. While some initial resistance had been expected from the industrialists on whom the tax
bore directly, the Treasury had anticipated that the City would
welcome the N.D.C. on the grounds that it would make the defence loan more attractive. This was not the case, however, because the bankers were concerned about the impact the measure
would have on industrial profits and the economic prosperity they
stimulated. The nature of Britain's recovery from the depression,
with the emphasis on domestic consumption and the diminishing
importance of international trade, had made the industrial and
financial communities far more interdependent than had traditionally been the case, and brought their interests more into line as
a consequence.[31] Indicative of the alliance between industry and
finance on this issue was the fact that Robert Horne, the Conservative MP, who spoke for the City, led the Conservative attack on
the N.D.C. in Parliament.

In their campaign to persuade the Treasury to withdraw or materially alter the N.D.C., business stressed that it constituted a
"tax on recovery." Because the tax was formulated to take a percentage of the increase in profits earned after the rearmament program had begun, the N.D.C. took a larger proportion of the profits of a firm that had only recently begun to operate in the black
than it did from a firm that had managed to maintain a consistent
level of profits in the early thirties. The Government countered
this objection by pointing out that every tax had certain inequities,
and that the amounts any firm would have to pay were so small
that they would not be crippling in any case.

On more specific grounds businessmen complained about the
difficulty of computing the tax, and the impossibility of estimating
it in advance in order to figure it into costs at the beginning of the
business year. To make pre-estimation impossible was, however,
precisely the Treasury's intention, for the revenue was supposed
to come out of a firm's profits, not be passed on to the consumer
in the form of higher prices. The complexity of the tax was the

[31] The high proportion of industrialists sitting on the boards of directors of Britain's banks tends to support this observation. Simon Haxey, *Tory M.P.*, p. 42.

4. "Capital Punishment"—Chamberlain offers business its choice of axes with which to cut its profits as John Simon looks on. The axes represent the base periods that businesses could choose in order to calculate their liability to the proposed N.D.C. *Daily Express*, May 26, 1937.

product of the Treasury's intention to touch only those profits resulting from the stimulus of the rearmament program, and its wish to make the tax as equitable as possible.

It was the Treasury's success in writing a tax that directly affected profits and could not be passed on that aroused the business community. Underlying the complaints about the complexities and inequities of the N.D.C. was a more fundamental concern about the precedent of the Government's assuming the power to compel business to relenquish part of its profits. It was business's resolute opposition to any form of government compulsion, not the comparatively trivial sums involved, that accounts for the strength of the reaction against the N.D.C. A year and a half earlier the same concern had moved industry to caution the Government about the manner in which it conducted the rearmament program.

In the weeks following the announcement of the N.D.C., the Treasury was beseiged by deputations representing various industrial groups wishing to express their objections to it. Chamberlain met with many of them personally and also gave a number of speeches in an effort to win support for the measure. Little support was forthcoming from any quarter, however. In Parliament the Labour opposition came down against the measure on the grounds that it was a ploy to gloss over and divert attention from the real problem that, it insisted, was profiteering. Labour did not, however, press its objections too hard, preferring instead to sit back and let the Government's own supporters attack the proposal. It must have come as little solace to Chamberlain when William Gallacher, the only Communist elected to the House of Commons, rose toward the end of an evening of almost continuous Conservative criticism of the N.D.C. and observed:

> We have had tonight a demonstration of a very interesting and instructive character. It has been very clearly demonstrated to the House and to the country that when profits are at issue, patriotism fades away. It is the great god Profit that is the one concern of the honourable Members opposite and the concern of the Chancellor of the Exchequer. . . . I ask the right honourable Gentleman to go ahead with his tax. I ask him not to soften

or to weaken, but to stiffen his plan, to defy big business in the City and to place the responsibility for Defence, which is the defence of profits, at the door of profits.[32]

After a month of trying to persuade business of the importance of the N.D.C. and of listening to its criticisms of it, the officials at the Treasury presented a very much scaled down and revised version of the plan to Parliament, one they believed met the significant criticisms without undermining the principles on which the tax was originally based. The new version, however, was almost summarily rejected by Robert Horne's group of Conservatives. As the correspondent for the *Guardian* noted: "No Chancellor of the Exchequer can feel comfortable in the face of opposition from that quarter. . . ."[33]

In the face of this continuing opposition Chamberlain was forced to withdraw the measure entirely on June 1, five days after he became Prime Minister. In the speech in which he announced his decision, Chamberlain stated that, although he felt "the business world has shown a certain amount of hysteria in criticising these proposals," which he continued to believe were sound, he had to consider

> what would be the result if we did force the Bill through this House and we still found industry held up, still found industry making no progress because it felt uncertainty, still found a general feeling that the tax was unfair.[34]

This concern about the economic effects of business resistance to the measure after it became law, combined with his reluctance to incur any more damage to his party's confidence in his leadership, persuaded him not to press the issue further.

Later in June a "new" N.D.C. was drawn up along the lines suggested by business. Taking a straight percentage of total profits, rather than focusing on the increase in profits that might have been attributable to the boom touched off by the rearmament program, the tax passed easily into law. As Robert Boothby wrote to

[32] *Parliamentary Debates* (House of Commons), Vol. 323, 4/27/37, col. 284-286.

[33] *The Manchester Guardian*, 5/25/37, p. 11.

[34] *Parliamentary Debates* (House of Commons), Vol. 324, 6/1/37, col. 924.

5. "Mousetrap."—Low's comment on the N.D.C. controversy when it was at its height. The new Chancellor of the Exchequer, John Simon, at the left appears distressed. *Evening Standard*, June 2, 1937.

Chamberlain, the one advantage of the original N.D.C. was that it made the business community very receptive to the new tax where previously it would have opposed any such tax at all. Only the Opposition objected to the new form of the N.D.C., pointing out that it would enable the businessman to calculate his tax ahead of time and pass the cost of it on to the consumer, leaving him with as large a profit as before.[35] In reformulating the tax, the Government surrendered the principle on which the N.D.C. was originally based, but managed to retain the revenue it was intended to accrue.[36] For Neville Chamberlain the affair was a humiliating public defeat that stemmed directly from his failure to appreciate the depth of the business community's antipathy to government intervention in the affairs of the private sector. The defeat provided him with an object lesson in the political pressure business could bring to bear, a lesson he did not soon forget.

THE RESPONSE TO THE DEFENCE LOAN

The Treasury chose to float the first installment of the National Defence Loan, £80 million worth, within two days of the announcement of the N.D.C. Amidst the furor surrounding that event, the loan received scant attention in the financial papers. There was, however, some speculation that the Treasury had chosen that time to launch the loan in order to take advantage of the flurry of demand for gilt-edged securities that was resulting from the uncertainty aroused in business circles by the N.D.C.[37] That there is no evidence in the Treasury files to substantiate this speculation by no means rules out the possibility.

What is interesting and revealing about the National Defence Loan is the response to it, especially on the part of the business community. While Labour condemned it, saying that it was no more justifiable to borrow for guns than it was to borrow to maintain the unemployed (for which Labour had been derided in 1931 by the business community, which demanded that financial orthodoxy be maintained), the business community accepted this

[35] Sabine, *British Budgets*, p. 113.

[36] T 171/336, 1937, Treasury file on the N.D.C.

[37] After the announcement of the N.D.C. the British stock market took its worst plunge since 1931.

breach of financial orthodoxy with relief. In a note to Chamberlain, Hopkins explained part of this apparent paradox, noting that:

> In financial quarters throughout the world (where perhaps more than anywhere else the credit of a nation is made or marred) borrowing for a current purpose which merely provides for men in idleness is regarded as one of the worst forms of borrowing, but borrowing for armaments (which is also theoretically impugnable) is regarded as unfortunate no doubt but still respectable.[38]

When faced with the choice of whether the Government's massive defence program should be financed through taxation or borrowing, few in the business community hesitated to choose the way they had earlier blasphemed as the short route to financial degradation. They recognized, and correctly, that to have undertaken to finance such a program through taxation would have torn the heart from the prosperity they had only begun to enjoy. A few diehards grumbled that cuts should be made in the dole, education, and the pay of civil servants, but even they recognized that the borrowing was necessary and infinitely preferable to a level of taxation that would have kept their integrity intact at the cost of their prosperity.

Despite their preference for the National Defence Loan, the business community did not overwhelm the Treasury with a patriotic rush for stock when it came to subscribing it. The issue was badly undersubscribed because its interest rates were not competitive in an expanding economy. Phillips, in an essay on the Government's borrowing problems, summed it up this way:

> Broadly speaking, prosperity, rising profits and rising commodity prices increase the supply of savings, but also make those savings more expensive to borrow and alter their distribution in such a way that industry tends to get all the money it wants while Governments and local authorities and building find it more difficult to raise money even at a somewhat higher rate of interest.[39]

[38] T 172/1853, undated, Hopkins to Chamberlain, replies to Labour attacks on borrowing.

[39] T 175/94, 7/28/37, Phillips on government borrowing.

The reason for this was that with profits running at a rate of 10 to 15 percent industry did not mind paying an extra percent or two to obtain the capital it required to expand its plant to take greater advantage of the boom. Governments and local authorities could not expect their revenue to expand in the same manner as industry could expect its profits to grow, and were severely handicapped by the increase in the cost of capital.

The public was aware that the National Defence Loan was undersubscribed, although the secret arrangement by which the Bank took up the unsubscribed issue made it impossible to determine by how much.[40] The loan's being undersubscribed was inflationary because the Bank issued paper money against the unsecured balance that it took over from the Treasury, that is, the Bank simply printed the money the Government was not able to borrow. The Treasury was not happy about the situation because it forced it to choose between limiting future borrowing or offering a higher rate of interest on future loans if it wished to prevent defence finance from becoming inflationary.

BALDWIN STEPS DOWN

Stanley Baldwin resigned as Prime Minister on May 28, 1937, after seeing the nation through the abdication crisis. Although Chamberlain had taken over most of the duties of the Prime Minister after Baldwin had fallen ill in the summer of 1936 and had prepared the way for the changes in administration and policy he felt were necessary, it was not until after he became Prime Minister in his own right that he implemented those changes. His accession to the leadership was not, however, welcomed by all Conservatives. One young progressive Tory remarked to a member of the press:

> Compared with Mr. Baldwin, Mr. Chamberlain is crude, harsh stuff and after seeing him fumble over his budget and practically abandon it before his critics—taxation of profits is the one thing Conservative M.P.'s really understand—we don't feel very happy about the new regime.[41]

[40] *The Daily Herald*, 4/30/37, pp. 1 and 14; *The Economist*, 5/1/37, p. 290.
[41] *The New Statesman*, "A London Diary," 5/14/37, p. 804.

157

Upon assuming his new office, Chamberlain's first act was to make some changes in the Cabinet that would allow him to pursue his policies. John Simon replaced Chamberlain as the Chancellor of the Exchequer, a position he told Chamberlain he approached "with humbleness" because he had "no special knowledge of national finance." Chamberlain replied that that "was not a bad qualification to start with."[42] In truth it was just the qualification Chamberlain sought, for he desired a man who would be wholly reliant on his Treasury advisers, and who would not deviate from the Treasury view of domestic and world affairs. That view, molded by his tenure of five and a half years at the Treasury, informed the policies Chamberlain was to advocate as Prime Minister, and he wanted the man who replaced him in the second most important position in the Cabinet to mirror it. In other changes, Duff Cooper, who had fought Chamberlain so long and hard over the role of the Army and the size of Army appropriations, was asked to move to the Admiralty. Chamberlain had seriously considered dropping Duff Cooper from the Cabinet entirely, but apparently decided it would be better having him in the Cabinet than asking embarrassing questions about the rearmament program from the Conservative backbenches in Parliament. Duff Cooper's replacement at the War Office was Leslie Hore-Belisha, who had shown considerable imagination and drive in his previous position as Minister of Transportation.[43] As we shall see, this appointment was somewhat deviously calculated. Anthony Eden, who had replaced Hoare as Secretary of State for Foreign Affairs in December of 1935, remained in that position. With this Cabinet Chamberlain was confident he could carry out the changes in the rearmament program he and the Treasury felt were vital.

[42] Martin Gilbert and Richard Gott, *The Appeasers* (Boston: Houghton Mifflin 1963), p. 53.

[43] R. J. Minney's *The Private Papers of Hore-Belisha* (London: Collins, 1960) gives a first-rate account of Hore-Belisha's years as Secretary of State for War.

The Rationing of the Services

FEARS ABOUT REARMAMENT

IN the course of 1936 the Treasury had become concerned that what they saw as "the system under which individual departments put up great new schemes at odd times, without regard to the aggregate bill," was "leading to financial chaos."[1] The Treasury believed that it had lost control of defence spending, and that as a result the whole defence program had lost any semblance of unified direction. The protean growth of the programs of the individual services was reflected in the appalling magnitude of the service estimates for the 1937 budget, which gave the actual impetus to the Treasury's decision that something had to be done.

The Treasury's concern was based on the belief that if the rearmament program continued on its course, the financing of it would cause serious dislocations in the economy that were bound to have profound social and political repercussions. It believed that the rearmament program was a far more profound threat to the social and political order than Germany was to the national security. At the heart of the problem was the location of revenue to pay for rearmament. The income tax, the primary source of tax revenue, was at its highest peacetime level, taking five shillings in every pound. Business had repeatedly expressed the desire that the tax be lowered to four shillings, voicing the fear that, if continued at its current level, the tax would bring the nation's prosperity to an early end. Phillips wrote that "taxation is already exceedingly high, so high that should any setback occur in our present position of general prosperity, the weight of taxation would exercise a hardening effect on enterprise."[2] Despite its high level,

---

[1] T 161/783/48434/02/1, 5/14/37, Hopkins essay on the future of defence finance.

[2] T 161/783/48431/3, 11/1/37, Treasury memorandum on raising capital to finance defence spending.

the tax was barely sufficient to cover the cost of maintaining and operating the services. It was universally agreed that any effort to rely on taxation to finance rearmament would lead the nation straight back into the depression.

Some members of the Government suggested that economies should be made in the social services to provide more funds for defence, but this suggestion was resisted by others who pointed out the risk of resulting social unrest. The Government recognized that to take from the poor in times of prosperity was to invite the kind of upheaval it so desperately wanted to avoid.[3]

As we have seen, borrowing was the only alternative, and it was one that was reluctantly embraced. Aside from having misgivings about embarking on any borrowing program, the Treasury was acutely aware of the limitations on the overall amount it could borrow before its actions would begin to have a deleterious effect on the economy. This outside limit was dictated by the classical economic belief that a nation could only borrow as much as its citizens saved. Any borrowing in excess of this limit would be tantamount to simply printing money, which would of course be inflationary. Edward Bridges explained Britain's situation in these terms:

> Owing to its shortage of native raw materials and foodstuffs, this country is particularly dependent upon imports which can only be paid for if the volume of our export trade is not impaired. This factor of the general balance of trade is closely connected with our credit. The amount of money which we can borrow without inflation is mainly dependent upon two factors: the savings of the country as a whole which are available for investment and the maintenance of confidence in our financial stability. But these savings would be reduced and our confidence at once weakened by any substantial disturbance of the general balance of trade. While if we were to borrow sums in excess of sums available for investment from savings, (by printing more paper pounds) the result would be inflation: i.e. a general rise in prices which would have an immediate effect on our export trade.[4]

[3] CAB 64/30, 11/12/37, minutes of Inskip committee on defence.
[4] T 161/855/48431/01/1, 12/8/37, Bridges on finance and defence.

Inflation's adverse effect on Britain's international credit position, and consequently on trade, was but one element of the Treasury's concern. Of tantamount importance was the belief that inflation would "lead to continuous pressure for increased wages, and produce conditions in which subversive ideas flourish."[5] Frederick Phillips spelled out the Government's fears about inflation resulting from excessive borrowing. In an analysis of the effects of inflation he wrote:

> The disadvantages of a great rise in prices are fairly obvious. It is gravely unjust to the owners of fixed incomes. The danger is not however so much from the rentiers as from the efforts of the salaried and wage earning classes to secure compensation for higher expenditure in the form of higher incomes. The rising cost of living is almost as potent a factor as unemployment itself in producing social unrest and providing a field for subversive propaganda. The legitimate grievance of the wage earner is intensified by the sight of the excessive profits which rising prices bring industry.[6]

He went on to note that while a certain amount of inflation is helpful to the Government, in allowing it to reduce the national debt cheaply, if prices begin to rise quickly, industrial profits will rise along with them, and wage earners will have little choice but to strike in order to maintain their standard of living. The catastrophic experience of Weimar Germany and more recently of France, with the social and political repercussions of inflation, served to underscore the Treasury's concerns, reinforcing its resolve that under no circumstances should Britain take such risks. At best such unrest would cut into arms production, and signal the dictators that Britain lacked the fiber to keep up in the arms race; at worst it would eventuate in the fall of the Government, the onset of socialism, and a consequent return to pacifism. Viewed in this light inflation was not only a threat to the national security, but to the whole social order. This was the apocalyptic vision that was at the heart of the Treasury's determination that defence borrowing, and the spending from which it derived, could not be allowed to exceed certain prescribed limits.

[5] T 161/783/48431/3,11/1/37, Treasury memorandum on raising capital for defence.

[6] T 175/94, 12/31/36, Phillips on inflation.

In the winter and early spring of 1937 the Treasury set about defining and determining those limits. Although the net national savings represented the theoretical outside borrowing limit, the Treasury decided initially that the limit should be calculated in terms of the size of the defence establishment the nation could afford to maintain out of tax revenue after the arms build-up had been completed. This limitation on the eventual size of the defence establishment automatically limited the amount to be spent on the arms build-up and, consequently, the amount to be borrowed to carry it out as well. The limitation was based on the rationale explained in Chapter II that recurrent expenditure cannot be allowed to exceed revenue. In a memorandum that was the central Treasury brief on the limitation of defence spending, Richard Hopkins explained:

> There is a grave danger that we may find at the end of the rearmament period that we have built up our armed forces to a level that is far beyond our capacity to maintain. There is no more certain way of drifting into bankruptcy than to borrow for a temporary capital purpose and then to continue borrowing—with no end in sight—for normal current requirements. [7]

He went on to argue that the services, therefore, had to be given concrete spending limits. In a refrain that was to be frequently heard he wrote:

> We must face the fact that we cannot do everything which we should like to do; that there is a limit to the amount of money that can be made available for defence measures and that the money available for defence must be allocated to those purposes, which, on a broad review of the whole situation are regarded as of prime importance.

The financial limit at which the Treasury arrived was £1,500 million, to be spent over the next five years. Of this figure £400 million was to be borrowed, and the rest raised out of revenue. The Treasury's basic assumption was that "we shall not, and cannot afford to, allow ourselves to slip quietly into American or French budgetary methods, but shall strive, at any rate till disaster over-

---

[7] T 161/783/48431/02/1, 5/14/37, Hopkins on the future of defence finance.

whelms us, to keep within the limits of decent finance." Warren Fisher summed it up very well in a handwritten note on the Hopkins paper in which he said: "We are rapidly drifting into chaos + are running the danger of undermining ourselves before the Boche feel it desirable to move."[8]

The Hopkins paper was the summation of a five-month Treasury study of the future of defence finance. Chamberlain had initiated the study in anticipation that its conclusions would provide the guidelines upon which his Government's rearmament policy would be based. His first important step after assuming office was to authorize the Treasury to draw up a plan for the complete re-evaluation of the rearmament program based on the Hopkins paper. The new Chancellor of the Exchequer presented the plan to the Cabinet at the end of June. The plan recommended that the defence ministries and Inskip submit to the Treasury estimates of the time required to complete the already authorized programs, their yearly cost, and the yearly maintenance cost of the products of the programs after their completion. The Treasury would then study the estimates and turn them over to the D.P.R., along with recommendations for the maximum it felt the nation could afford to spend. It would be up to the D.P.R. to order the priorities among the programs, deciding in light of both the financial and international situations which programs should be carried on and which should be terminated, as well as the amount each service should be allotted. The Treasury also recommended that, until this review was completed, decisions on new defence programs should be postponed.[9] Thus was order to be imposed on the chaos of the rearmament program.

The response of the service ministers to the Treasury plan was in many ways predictable. Swinton and Duff Cooper as well as Inskip expressed concern lest the review would hold up the completion of programs already under way. Simon explained that programs that had already been approved by the Cabinet and authorized in detail by the Treasury would not be affected by the review, hence

[8] T 161/783/48431/02/1, 5/14/37, Fisher note on the Hopkins paper, dated 6/1/37.

[9] CAB 24/270, CP 165 (37), 6/25/37, John Simon on defence spending and the Exchequer.

contracts that had already been let would not be interfered with. However, he hastened to point out that all programs that had been approved by the Cabinet (or the C.I.D. or the D.P.R.) in principle, but not by the Treasury in detail, were to fall within the scope of the review. The service hit hardest by this stipulation was the Army. The compromise solution to the question of the equipment of the T. A., for which Duff Cooper had struggled so long, had been agreed to by the Cabinet little more than a month earlier, and the War Office had not had time to draw up detailed plans to submit for Treasury ratification. Therefore the T. A. plan fell within the scope of the review, and no action could be taken on it until that review was completed. It was a bitter blow, which the new Secretary of State for War, Hore-Belisha, protested to no avail. In response to Hore-Belisha's arguments that the suspension of the program would set the Army's rearmament program hopelessly behind, destroy any chance of the creation of industrial capacity to meet its needs for war materiel, and discourage contractors from taking Army contracts at all, Simon simply stated that the sum required to implement the program was "out of the question."[10]

INSKIP'S REVIEW OF DEFENCE EXPENDITURE

In the course of the summer, while the services were preparing their estimates as specified by the Treasury paper, a basic change was made in the procedure for reviewing those estimates. The Cabinet decided that in view of their other responsibilities, the members of the D.P.R., who were all cabinet ministers, could not afford to spend the time a thorough review of the defence program would require, and assigned Thomas Inskip to review all the relevant material with the aid of a small committee of select civil servants. Upon completion of this review Inskip was to present a report to the Cabinet containing his conclusions and recommendations on the direction and shape that rearmament should assume. Horace Wilson, Chamberlain's Chief Industrial Adviser and close confidant, Arthur Robinson, the head of the Supply Board, and Maurice Hankey comprised what came to be known as the Inskip Committee. Beginning in October they met with representatives

[10] CAB 16/181, 3rd meeting of D. P. (P), 7/17/37.

from the services, the Treasury, and the Foreign Office to hear the opinions of the various departments on defence questions, and to try and evolve a coherent policy based on the realities of the situation. In the course of their review the committee was in close consultation with the Treasury, which, beside being responsible for going over the figures submitted by the services, had an abiding interest in the outcome of the review. Of those with whom Inskip consulted closely, both Hankey and Robinson put their concern for Britain's military security ahead of that for her financial security. Despite their predilection, however, both men had to accept the premise that the direction of defence spending had to be altered. It was, after all, the premise that had brought the committee into existence.

Inskip accepted the primacy of keeping defence spending within the limits prescribed by the Treasury; however, he could not accept limits that would jeopardize the security of the nation. In a note to the Treasury dealing with the global amount to be spent on defence over the next five years, Inskip stated his view in these terms.

> It [the total to be spent on defence] is . . . largely, if not entirely a question for the Treasury. I realise that the figure must be fixed with reference to the military situation—needs must when the devil drives—but except in circumstances that leave us with no alternative, I suggest that expenditure should not be contemplated on a scale which is likely to exhaust our financial resources. The question is how are we to reconcile the two desidirata, first, to be safe, secondly, to be solvent. It is necessary to form some clear conception as to the degree of safety at which we should aim.[11]

The reconciliation of safety with solvency was a task of considerable difficulty. The real crux of Inskip's problem was to arrive at a solution that would provide the nation with the highest degree of safety possible within the financial limits set by the Treasury.

In the course of his review Inskip issued an interim report to the Cabinet on Defence Expenditure in Future Years in December 1937, and a final report the following February. The interim re-

[11] T 161/855/48431/01/1, 11/23/37, Inskip to the Treasury on the future of defence expenditure.

port provided a general analysis of the finance and defence situations and proposed some tentative recommendations concerning the programs of the services. The final report was detailed and specific, basing its proposals on an analysis of costs and risks, and giving concrete recommendations concerning particular programs. While the interim report was written in collaboration with the Treasury, the final report was the product of Inskip's own synthesis.

THE RATIONALE FOR RATIONING

In the interim report the first section, which was written at the Treasury, was concerned with explaining why it was vital to national security to limit defence expenditure. Although the Treasury's concern with the maintenance of economic and social stability was the motivating force underlying its insistence on the limitation of spending and borrowing for defence, it felt it had to justify that limitation as a positive force in the emergent rearmament policy if the Cabinet were to be fully persuaded of its necessity. The central argument was that:

> Security is more than armaments. Military effort itself depends directly on our resources in manpower and industry, but financial resources and economic strength more generally are essential components in the defence structure.[12]

Britain's economic and social stability was represented as a deterrent to her enemies.

> Nothing operates more strongly to deter a potential aggressor from attacking this country than our stability, and the power which this nation has so often shown of overcoming its difficulties without violent change and without damage to its inherent strength. This reputation stands us in good stead, and causes other countries to rate our powers of resistance at something far more formidable than is implied merely by the number of men of war, aeroplanes, and battalions we should have at our disposal immediately on the outbreak of war. But were other countries to detect in us signs of strain, this deterrent would at once be lost.[13]

[12] CAB 64/30, first meeting of Inskip's committee, 10/28/37.
[13] CAB 24/273, CP 316 (37), 12/15/37, Inskip's Interim Report on Defence Expenditure in Future Years.

At the heart of the Treasury's case, however, was the doctrine that finance was the fourth branch of the armed services. Under the precepts of this doctrine the maintenance of economic stability becomes one of the cornerstones of the nation's defence structure, and an integral part of the defensive strategy. It is represented in the interim report in the following terms:

> The maintenance of credit facilities and our general balance of trade are of vital importance not merely from the point of view of our strength in peacetime, but equally for purposes of war. This country cannot hope to win a war against a major Power by a sudden knockout blow; on the contrary, for success we must contemplate a long war in the course of which we should have to mobilise all our resources and those of the Dominions and other countries overseas. . . . Germany is likely to be the aggressor and will endeavour "to exploit her superior preparedness by trying to knock out Great Britain rapidly or to knock out France rapidly, for she is not well placed for a long war in which Sea Powers, as in the past are likely to have the advantage." We must, therefore, confront our enemies with the risks of a long war, which they cannot face. If we are to emerge victoriously from such a war, it is essential that we should enter it with sufficient economic strength to enable us to make the fullest use of resources overseas, and to withstand the strain. While, therefore, it is true that the extent of our resources imposes limitations upon the size of the defence programmes which we are able to undertake, this is only one aspect of the matter. Seen in its true perspective, the maintenance of our economic stability would more accurately be described as an essential element in our defensive strength: one which can properly be regarded as a fourth arm in defence, alongside the three services without which purely military effort would be of no avail.[14]

By maintaining the stability of the pound and Britain's international credit position, Inskip felt that, in the event of war, Britain's credit could be used to purchase arms abroad in the interim between the outbreak of the war and the time when her industrial capacity could be converted to arms production. Of course arms

[14] Ibid.

were not the only thing Britain had to rely on credit to purchase. She was dependent on the importation of food, and in wartime, when she was not producing for export, she could only make such purchases if her credit were good. Dislocating her economy by turning too large a proportion of it over to arms production in peacetime would hurt her trade balance and credit by diminishing the amount she produced for export. Here again was the problem of balancing Britain's security against her international solvency.

The strategy of limiting Britain's military preparations to those sufficient to protect her from a sudden knockout blow was the Treasury's and Inskip's answer to that problem. By concentrating on the defence of the home islands, they calculated, the greatest amount of security could be obtained for the least cost. Equally important was the consideration that the production of purely defensive weaponry would cause a minimum of dislocation in the industrial sector.

It was Maurice Hankey who proposed the order of Britain's priorities upon which this strategy was based. He wrote to Inskip:

> First priority must be given to the security of the United Kingdom and its Capitol, which constitute the central keep and base of the whole system of Imperial Defence. To quote the words of Earl Balfour in 1905, "National independence must not be staked—even in appearance—on too nice a calculation of chances. We cannot sleep secure with only a bare margin of probability in our favour."[15]

The second priority was to keep the nation's lines of communication, that is, the sea lanes, open, while the third was to defend British territories.[16] "Our fourth objective, which can only be

---

[15] T 161/855/48431/01/1, 11/23/37, Hankey to Inskip.

[16] In ranking the defence of Britain's territories, which is to say the Empire, as the nation's third priority, Hankey and Inskip gave a clear indication of the Government's view of the Empire's importance in the defence scheme. The reluctance of most of the Commonwealth nations to come to Britain's aid if she were to become involved in a war on the Continent, along with their military vulnerability and reliance on the British Navy for defence, made them distinct liabilities. On the other hand, the Commonwealth nations were Britain's primary source of food and raw materials, a consideration of tantamount importance in the event of a major conflict. In terms of Britain's immediate survival after the outbreak of war, however, their defence was third in the Government's order of priorities. For a discussion of Imperial relations during the period see D. C. Watt, *Personalities and*

provided for after the other objectives have been met, is co-operation in the defence of the territories of any allies we may have in war."[17] These were the priorities that Inskip set forth in the interim report, and they had important implications for both the Army and the R.A.F.

RATIONING AND THE ARMY

The Army, which had come to be called the Cinderella Service, was the most profoundly affected. The relegation of cooperation with Britain's allies to the lowest priority, conditional on the other needs being met, augured the end of the Army's Continental role. In light of the Prime Minister's frequently expressed doubts about the importance of that role, this is not surprising. The Treasury believed that if the primary role of the Regular Army became the defence of imperial commitments and anti-aircraft defence at home, substantial savings could be effected due to the elimination of the need for the costly weapons required for war on the Continent. The Army of course could not be expected to be pleased with this change. The General Staff was completely oriented toward war on the Continent. It was the arena of the world's most sophisticated warfare, requiring the most advanced weapons and the most refined strategy. To deny the professional soldier access to the battlefield that he had dedicated his career to being successful on, and to offer him police work in the colonies and service as an auxiliary to the R.A.F. at home, was a cruel blow indeed. The Army was in no condition to protest, however. During this period Hore-Belisha was in the midst of what may be described as a purge of the General Staff, hence he was not in a position to contest the change in the Army's role. There is every indication that Chamberlain, who had more than once criticized the General Staff's lack of drive, imagination, and talent, encouraged Hore-Belisha to conduct the shakeup. He noted in a letter written the first of August that Hore-Belisha was

---

Policies: Studies in the Formulation of British Foreign Policy in the Twentieth Century (London: Longmans, 1965) and Nicholas Mansergh, *The Commonwealth Experience* (New York: Praeger, 1969).

[17] CAB 24/273, CP 316 (37), 12/15/37, Inskip's interim report.

doing what I put him there for and has already stirred the old dry bones up till they fairly rattle. Things are even worse at the War Office than I feared and I foresee that I am going to have a fierce struggle there before we can settle down to rebuild on sounder foundations.[18]

This upheaval in the highest ranks of the Army resulted in a split between the Secretary of State and the General Staff on whom he relied for the preparation of the Army case that he was responsible for presenting to the Cabinet. Divorced as Hore-Belisha was from the General Staff (his closest adviser on military questions at this time was the military correspondent for the *Times*, Basil Liddell Hart),[19] he was in no position to hear, much less argue its case. The result was that, during the first nine months of his tenure at the War Office, Hore-Belisha was virtually ineffective in defending the Army's interests in the Cabinet. In those nine months it was stripped of what it considered its most important role. The actual recommendation of the interim report was that the Army's continental role be eliminated. The reasons given were that France no longer expected Britain to contribute a large continental force in their alliance, that the recent German guarantee of Belgian neutrality eliminated the need for a British force to secure that area, and that the demands for the presence of the Army in the Empire were of a higher priority than its commitment to the Continent. The unstated, but dominant reason, of course, was cost.

THE BOMBER CHALLENGED

In the course of Inskip's review in the fall of 1937, the R.A.F.'s bomber strategy came under close scrutiny due to its burgeoning expense. Although its proposal for Scheme H had been denied by the Cabinet in February on the grounds of cost, by June the Air

[18] Keith Middlemas, *Diplomacy of Illusion: The British Government and Germany, 1937-39* (London: Weidenfield and Nicolson, 1972), p. 120. This study was the first to consider the impact rearmament had on British foreign policy in the thirties.

[19] Liddell Hart had advised several Secretaries of State for War, and was in fact introduced to Hore-Belisha by Duff Cooper, his predecessor. For the relationship between Liddell Hart and Hore-Belisha see Liddell Hart's *Memoirs*, Vol. I, and Minney's *The Private Papers of Hore-Belisha*, pp. 54-55 and *passim*.

6. "New Broom at the War Office"—Hore-Belisha shakes up the War Office while the Colonel Blimps look on aghast. *Evening Standard*, December 8, 1937.

Ministry had begun to circulate a new and even more expensive proposal, Scheme J, which envisaged an increase of twenty-two bomber and eight fighter squadrons over the currently authorized plan. The Treasury was skeptical of the need for that many additional bombers, and hostile to the proposal in general as it threatened to interfere with Inskip's review. At a June C.I.D. meeting Swinton was questioned about the advisability of the Air Ministry's bomber strategy. He replied that it "was strategically most undesirable that fighter strength should be increased at the expense of our bomber strength."[20] This proved to be the opening volley in a battle between the Treasury and the Air Ministry that was to continue through the outbreak of the war. The debate began to build as it became evident that the work Sir Henry Tizard and Robert Watson-Watt had been doing on a top secret project referred to as radio direction finding, known today as radar, would enable the fighter command to meet and repulse bomber attacks against England effectively, whereas the presumption had previously been that "the bomber would always get through."[21] This knowledge prompted some thinking at the Treasury. In late October Bridges wrote in a note on air defence, "The general impression which one gathers is that defence against bombing shows signs of catching up against attack."[22] The thought was enlarged upon, and on November 4 Inskip wrote to Swinton asking for an estimate of the savings that could be realized if the bomber force were to be reduced in size and readiness. The request was not popular at the Air Ministry, for it flew straight in the face of the Air Staff's philosophy that "we must give an enemy as much as he can give us."[23] Swinton shot back a reply the same day that detailed at length the Air Ministry's case for the bomber. The basic premise was that to deter an enemy from attacking, one's own offensive threat must be seen by the enemy to be as effective as his own. If one relaxes one's offensive threat, the enemy can ease his defensive preparations and concentrate on attack. To this argument he added that, although the radio direction finding tests

[20] CAB 2/6, 294 (8), 6/17/37.

[21] For an account of the development of radar see Ronald W. Clark's biography, *Tizard* (London: Methuen, 1965), especially chapters 4 and 5.

[22] T 161/778/41764, 10/26/35, Treasury notes on air defence.

[23] CAB 64/60, D.R.C. Review Committee, 11/2/37.

were promising, "it would be an illusion to suggest that we have a sure means of defence. Counter-attack still remains our chief deterrent and defence."[24] He concluded by noting that, in view of the public pledges made by the Government about its resolve to maintain air parity with Germany, a shift away from the policy of parity in striking capability might prove politically embarrassing.

Despite the Air Ministry's emphatic insistence on the necessity of the bomber deterrent, neither Inskip nor the Treasury was convinced. Inskip wrote, "For myself I do not accept the dictum of the Chief of the Air Staff that we must give the enemy as much as he gives us."[25] The fighter's attraction was that it could be built more cheaply and faster than the bomber, and required less manpower to fly and maintain. At the December 2 meeting of his committee Inskip reiterated his view that the best air policy for Britain might be to concentrate on fighters and defence. Hankey and Fisher took up the idea and integrated it into the priorities that had been agreed on. Fighters would be necessary to avert the knockout blow, it was argued, and once that was survived the nation could turn its resources to the creation of its own bomber force. If war never came, the nation would be spared the massive cost of the bomber program. This was the logic behind the defensive role Inskip's interim review suggested for the R.A.F. Inskip proposed that the bomber increases suggested in Scheme J be eliminated, and that fighter increases be accorded first priority, "given that our main object is the defence of this country."[26] Unlike the Army, the Air Ministry was fully prepared to contest the issue with all the techniques at its command.

Only the Admiralty found its role in the national defence unchanged by the interim report. While acknowledging the validity of the Navy's argument for the two power standard, Inskip felt constrained by the financial situation to recommend only that their existing building program be completed as authorized.

THE LOGIC OF APPEASEMENT

While the services contended with the Treasury over the roles assigned them in the national defence strategy outlined in the

[24] CAB 64/30, 11/26/37, Swinton to Inskip.
[25] T 161/855/48431/01/1, 11/23/37, Inskip to Treasury.
[26] CAB 24/273, CP 316 (37), 12/15/37, Inskip's interim report.

interim report, the Foreign Office was engaged in a bitter struggle with the Treasury over the role that foreign policy was to play in that strategy. In the course of Chamberlain's last year as Chancellor of the Exchequer he and his advisers at the Treasury had come to the conclusion that if the rearmament program continued to expand at the rate it was then expanding, there was a profound possibility that it would undermine and destroy the existing economic and social structures. With the decision to limit rearmament the crucial question arose of "what must be the guiding lines of our foreign policy in light of the defence forces which we can afford to maintain permanently?" The Treasury was quick to realize that "ultimately our foreign policy must largely be determined by the limits of our effective strength."[27] Its conclusion was that, where the nation's effective strength was too limited to affect an issue that concerned it, conciliation or, as it quickly came to be called, appeasement, had to be used instead of the threat of force. Appeasement became necessary because the Treasury believed, and the Government concurred, that the British economy could not stand the strain that the marshalling of force sufficient to meet Britain's commitments would entail. The battle between the Foreign Office and the Treasury centered on the question of whether appeasement was an effective substitute for force in the international arena.

By late 1937 the two departments were in agreement on one undeniable fact: Britain was in a weak military position and that position was deteriorating relative to that of her enemies. A C.O.S. review of Britain's strength made that fact painfully clear by pointing out that, although Germany's army was not yet strong enough to achieve a quick victory over France, and her air force was still too weak to knock out either France or Britain, she was moving quickly towards a level of strength at which she would be capable of both. While Germany's military strength was the most immediate problem, it was by no means the only one. Japan remained a dangerous imponderable in the Far East, and Italy's increased hostility posed a profound threat to Britain's lines of communications with the Empire. Sitting astride the sea routes through the Mediterranean to Suez, Italy was ideally situated to

[27] T 161/783/48431/02/2, 7/19/37, Bridges note.

deal a serious blow to British shipping. The C.O.S.'s conclusion was that Britain's military was insufficient to meet her combined defensive commitments. The subcommittee insisted that, "We cannot, therefore, exaggerate the importance, from the point of view of Imperial defence, of any political or international action that can be taken to reduce the number of potential enemies and to gain the support of potential allies."[28]

In light of this situation the Foreign Office felt that the proper course was to strengthen Britain's ties with her allies, especially France, and endeavor to persuade the United States to make some kind of commitment against fascism, in order to confront Britain's enemies with as unified an opposition as possible. Meanwhile, it felt that relations with the dictators should be correct and firm in an effort to diffuse potential conflicts. The Foreign Office emphasized that until Britain developed her military strength to the point that she could contend on equal terms with her enemies, the possibility of war was constant.

To the Treasury and to Chamberlain this position was unacceptable. As they had stated many times, Britain could not afford to match her enemies in armaments. Conciliation was the only course they could see by which it was possible to save the nation from the threat posed from without by too few arms and from within by too many. Consequently both Chamberlain and the Treasury took exception to the Foreign Office's handling of relations with Germany and Italy.

In the case of Germany their differences with the Foreign Office centered more on attitude than on policy. Hitler had raised the possibility of Britain's offering Germany colonial concessions in Africa in discussions with Simon and Eden in early 1935, and the Foreign Office had discussed the issue sporadically with the Germans on an unofficial basis. By 1937, however, the Foreign Office had concluded that any such territorial concessions should only be considered as part of a plan to buy time to rearm. Chamberlain and the Treasury on the other hand were far more sanguine about the possibilities of such concessions, believing that they could form the basis of a comprehensive settlement that would

[28] CAB 4/26, 1366-B, 11/12/37, C.O.S. report on Britain's strength with respect to her potential enemies.

adequately compensate Germany for the territory in Eastern Europe that she was insisting should be returned to the Reich. They felt that if a settlement could be achieved quickly, there was a good chance that European tranquility could be re-established, and that the rearmament program could thus be deferred. Led by Eden and Vansittart, the Foreign Office was more sceptical about Germany's intentions, and chose to proceed warily in its negotiations with Hitler in 1937.[29]

The Treasury fumed over the Foreign Offices's dilatory and, to its mind, overcautious behavior, characterizing Eden's policy as a negative one intended to do nothing more than "keep Germany guessing." Because it effectively foreclosed any possibility of achieving a settlement, such a policy, the Treasury believed, was wholly inappropriate to the situation in 1937. The Treasury argued that in the conduct of its foreign policy the Government had to assume:

> that the Nazi struggle is primarily one for self-respect, a natural reaction against the ostracism that followed the war; that its military manifestations are no more than an expression of the military German temperament (just as our temperament expresses itself in terms of sport); that Hitler's desire for friendship with England is perfectly genuine and still widely shared; and that the German is appealing to the least unfriendly boy in the school to release him from the Coventry to which he was sent after the war, . . . [even if they believed] that Germany is a beast of prey waiting to pounce; that the first victim will be, say, Czechoslovakia; that our turn will come; and that Hitler's professions of friendship are so much dust thrown in our eyes.

The Treasury contended that, if Britain acted upon the second set of presumptions, while in a militarily inferior position, war was inevitable; whereas, if the actions of the Government were in

[29] This brief description of the African colonies question is drawn primarily from Keith Middlemas's *Diplomacy of Illusion*. See also C. A. MacDonald, "Economic Appeasement and the German 'Moderates,' 1937-1939," *Past and Present*, Vol. 56 (August 1972). The definitive work on economic appeasement is Bernd Jürgen Wendt's *Economic Appeasement: Handel und Finanz in der britischen Deutschland-Politik, 1933-1939* (Düsseldorf: Bertelsmann Universitätsverlag, 1971).

keeping with the first set of presumptions, even if they were not believed, Britain would at best maintain the possibility of peace and at worst have the longest possible time to rearm. The Treasury used the following analogy to clarify the rationale behind this line of thinking:

> It is a well known maxim at Bridge that if there is only one distribution of the opponents' cards which will enable you to make your contract, your play should be based on that distribution, *whether probable or not*, as the only play that affords a prospect of success.[30]

This was not the approach that the Foreign Office was pursuing, and the Treasury was openly critical. A memorandum written in response to a Foreign Office paper on the conduct of British foreign policy outlines the Treasury's view of the relationship Britain should try to cultivate with Germany. It begins by stating that "we should seek to dispel the atmosphere of suspicion which surrounds our relationship with them [the Dictator Powers] (particularly with Germany), and makes it necessary to regard them as potential foes." The Treasury felt that the Foreign Office position, "while it allows us to continue with the talks and discussions, seems to bar out indefinitely any real forward move on our part, such as is needed to break down the vicious circle of mutual suspicion, and bring about a general appeasement." Turning to what it saw as the Foreign Office's attitude toward Germany, the Treasury rhetorically asked:

> Is not it most unfair to regard Germany in the same sense as the other two Powers as "an aggressive power" whose "aims are inimical to British interests"? . . . For all the bitterness left by the war and the not always fortunate treatment of German minorities under the Treaty of Versailles, Germany has committed no aggression since the war unless the reoccupation of her own territory can be so described. Germany's only aim that can be described as "inimical" to this country is the possession of some colonial territory, but this in itself should not be a hopeless barrier to friendship.

[30] T 172/1801, 8/10/37, Treasury critique of the Foreign Office's conduct of policy with Germany, written by Edward Hale.

Having thus dispelled any qualms that Germany's rather belligerent posture might have caused, the paper concluded:

> It is therefore much to be hoped that we may now adopt a less negative policy, and above all a more friendly attitude towards Germany. . . . If this friendly feeling with Germany can be promoted there should be a real chance of obtaining a settlement of most of the major troubles which beset us, and of obtaining relief from the intolerable and growing burden of armaments.[31]

This final sentence explicitly expresses the relationship between the policy of appeasement and the Treasury's belief in the need to curb rearmament.

The disdain of the Treasury secretaries for the Foreign Office's policy position can be seen in the notes they added by hand to this paper. J.A.N. Barlow, an under-secretary, wrote:

> Germany is always regarded as our most serious menace. Is she not also our best hope? Yet the F. O. contemplate merely a continuance of pin-pricks, of "preventing" her from attaining her ends.
>
> The practical conclusion of the Foreign Office's paper is merely larger armaments. It betrays no recognition of the economic aspect or of its value as a weapon for peace.[32]

It was left to Warren Fisher to convey bluntly the Treasury's view of the Foreign Office's performance.

> Our international policy since the Armistice has been dictated by France unto our own Foreign Office. We can see the results. Unless the Cabinet impose on the F. O. a reversal of past policy and attitude, we shall be landed sooner or later in disaster.[33]

While the differences between the Foreign Office and the Treasury over the tenor of Britain's diplomatic attitude toward Germany were considerable, it was over Italy that the conflict developed that eventuated in Eden's resignation and the first Cabinet crisis of Chamberlain's premiership. All sides recognized that

---

[31] T 161/779/41815, 12/1/37, memorandum by Edward Hale on the Foreign Office's view of Britain in the world.

[32] Ibid., Barlow's note at the bottom.

[33] Ibid., Fisher's note at the bottom.

Italy offered the best opportunity to realize the C.O.S.'s fervently expressed hope that the number of Britain's potential enemies could be diminished through the skillful conduct of diplomacy. The question was how that end could best be achieved.

Although Anglo-Italian relations had deteriorated seriously in the aftermath of the Hoare-Laval Affair and had been further exacerbated by Italy's blatant intervention in the Spanish Civil War, by the summer of 1937 Mussolini had begun to give indications that he would not be averse to some kind of accord with Britain. Increasingly uneasy about the strategic implications for Italy of Hitler's plan to absorb Austria into the Reich, he was canvassing his alternatives in case his fears about the intentions of his erstwhile comrade in arms were borne out. In his initiatives to the British he made the recognition of Italy's conquests in Ethiopia the one condition for rapprochement.

This condition sharply divided the Foreign Office and Chamberlain. At the Foreign Office Eden and Vansittart felt that it would be not only unethical but dangerous to sanction Italy's transgressions in Africa without requiring in return a concrete gesture of good will; the withdrawal of some Italian "volunteers" from the civil war in Spain came quickly to their minds. They contended that if Mussolini was unwilling to make such a gesture it would indicate that he was not serious about rapprochement, and that he would be free to repudiate at his own convenience any agreement stemming from it. In short they suspected that Italy was trying to secure recognition of her Ethiopian conquests in return for evanescent promises of friendship and good will that would cost her nothing and could easily be broken.

Chamberlain's perspective on the issue was quite different. To him the elimination of Italy from the lineup of Britain's potential enemies was not an option but an imperative. Concerned that the time for securing Italian friendship was fast running out and knowing that Hitler was pressing Mussolini for an alliance (it was known that Hitler was planning to visit Mussolini in May of 1938),[34] Chamberlain had little patience with the Foreign Office's suspicions and compunctions. He wrote in his diary on September 12, 1937:

[34] Roskill, *Hankey*, Vol. III, p. 303.

I am terribly afraid lest we should let the Italian situation slip back to where it was before I intervened. The F. O. persist in seeing Mussolini only as a sort of Machiavelli putting on a false mask of friendship in order to further nefarious ambition. If we treat him like that we shall get nowhere with him and we shall have to pay for our mistrust by appallingly costly defences in the Mediterranean.[35]

While he may have harbored some suspicions about Mussolini's intentions, Chamberlain was far more worried about the consequences of alienating Italy by pressing for concessions that would be of no benefit to Britain's security.

Although Eden and the Foreign Office may have had reservations about Chamberlain's and the Government's Italian policy, it was their responsibility to implement it. Chamberlain found, however, that their efforts to discharge that responsibility left much to be desired. Their apparent inability or reluctance to set aside their reservations and vigorously pursue what Chamberlain and his Cabinet believed to be the main chance moved him to establish his own lines of communication with the Italian government. Using his late half-brother Austin's widow, Lady Ivy Chamberlain, and an American, Joseph Ball, as intermediaries, he endeavored to circumvent the Foreign Office and maintain Italian enthusiasm for an agreement.

These personal diplomatic initiatives, which were taken without the knowledge of the Foreign Office, were pursued in large part out of a desire to avoid a confrontation with Eden in the Cabinet on the issue of foreign policy. Chamberlain was well aware that such a confrontation could have serious political repercussions within both the Government and the country at large because of the widespread popularity and prestige Eden had gained as a result of his staunch support of the League. A confrontation with Eden had the potential to touch off the kind of political crisis that Chamberlain wished to avoid early in his premiership.

In January 1938 Chamberlain attempted to bring the Foreign Office more under his control without directly affronting Eden by "promoting" Vansittart from his position as Permanent Under-

[35] Middlemas, *The Diplomacy of Illusion*, p.132.

Secretary of State. Vansittart, who was Eden's closest adviser as well as the leading source of opposition to Chamberlain's views on foreign policy in the Foreign Office, was made Chief Diplomatic Adviser, a largely honorific title, without specific powers or responsibilities, that was created for the occasion. This removed the day to day conduct of the Foreign Office's affairs from Vansittart's hands, and effectively eliminated his strong views on Germany, Italy, and rearmament from the committees involved in the formulation of the Government's foreign and rearmament policies. Although the new Permanent Under-Secretary, Alexander Cadogan, was the foremost member of the group at the Foreign Office that supported the Prime Minister's views, there was little he could do to alter Eden's continued opposition to Chamberlain's Italian policy.

The conflict between the two reached the crisis point in February as Hitler began to apply increasing pressure on Austria and Mussolini became more insistent in his pursuit of British support. Chamberlain and his Foreign Secretary were in sharp disagreement over how to proceed, and in the course of a particularly animated discussion of the matter, Chamberlain brought his private initiatives to Eden's attention. This evidence of the Prime Minister's lack of confidence in his conduct of affairs brought home to Eden the extent of their differences, moving him to resign on February 20, 1938.

Although Eden's resignation shook the Government, it did not result in any debate on the merits of the issue in the Cabinet, where a substantial majority supported Chamberlain's position, in Parliament, or in the press. Discussion of the matter was limited to personalities, as Eden chose not to broaden it to involve questions of policy. Consequently the political stability of Chamberlain's Government was not seriously weakened.[36]

At the heart of the differences between Chamberlain and the Foreign Office was the unwillingness of Eden and Vansittart to accept either the decision to limit rearmament or its implications.

[36] For the details of the events leading to Eden's resignation see David Dilks, *The Diaries of Sir Alexander Cadogan, 1938-1945* (New York: G. P. Putnam's, 1972), pp. 27-55; Middlemas, *The Diplomacy of Illusion*, pp. 128-156; and Feiling, *Chamberlain*, pp. 337-340.

Although it had been made clear that the decision meant that Britain did not presently and would not in the foreseeable future have the capacity to defend herself against Japan, Italy, and Germany simultaneously, Eden and Vansittart persisted in pursuing a foreign policy based on the assumption that the rearmament program would give the nation that capacity. In doing so they further increased the risks to which Britain was then exposed.

Eden's successor as Foreign Secretary was Lord Halifax, former Viceroy to India and a close friend and confidant of Chamberlain's. Once he took office, with Cadogan as his Permanent Under-Secretary, the Foreign Office quickly fell into line. From that point on the Government's policy of appeasement was conducted through the Foreign Office, which implemented it without demur. The focus of the Foreign Office's attention in the months that followed was on Germany rather than Italy, however, as whatever opportunity there might have been to secure Italian friendship evaporated with Hitler's occupation of Austria little more than three weeks after Eden's resignation.

The views of the Foreign Office notwithstanding, Inskip's interim review tacitly recognized the weakness that the Treasury's restrictions on rearmament imposed on the nation's defences, and emphasized the importance of achieving the foreign policy goals that they advocated. He wrote:

> in the long run the provision of adequate defences within the means at our disposal will only be achieved when our long-term foreign policy has succeeded in changing the present assumptions as to our potential enemies. . . . This [Britain's rearmament] policy must be based on the existing international situation, though it should take into account the changes in that situation which it is reasonable to expect, and those changes which it should be the aim of our foreign policy tó bring about.[37]

The proposals set forth in the interim review reflected the relationships among finance, rearmament, and diplomacy that the Treasury saw as necessary to the national survival. Inskip, however,

[37] CAB 24/273, CP 316 (37), 12/15/37, Inskip's interim report.

had doubts as to the wisdom of the plan. He was haunted by the truth of Hankey's observation that:

> It is clear that, unless something can be done in the field of foreign policy to reduce our foreign commitments, risks have to be taken somewhere, either in the field of finance or that of defence.[38]

Although the general outlines of the Government's rearmament policy had been sketched, Inskip had yet to determine the final balance in pounds and pence. In the striking of this balance the decision was to be made as to where the risks were to be taken.

THE RESPONSE TO INSKIP'S INTERIM REPORT

When the interim report was finally issued on December 15, 1937, it was not greeted with great enthusiasm, even by the Treasury, which had such a prominent hand in its outcome. On the 17th Richard Hopkins wrote that "If . . . the Cabinet approve these conceptions [estimated to cost £1,500 million] and in February they are proved to cost £1,700 million, how are we to claw back (Mr. Churchill's favourite phrase) £200 million out of the dog's mouth?"[39] Edward Bridges expressed his displeasure with the defence departments' reception of the financial arguments in these terms:

> I have been amazed at the attitude of cynical skepticism with which the statement [on finance] . . . is regarded in certain quarters. There are many people, whom one would have thought knew better, who . . . seem to regard any attempt to ensure that this country lives within its means as about the order of importance as not leaving a party without saying good-bye to one's hostess, i.e. something which isn't done in the best circles, but which would not really have any effect on how things went in the great world the next day. I think the Chancellor must be prepared to deal with the suggestion that, if we decided to disregard the canons of sound finance we should be able to

[38] T 161/855/48431/01/1, 11/23/37, Hankey to Inskip.
[39] T 161/855/48431/01/2, 12/17/37, Hopkins memorandum concerning Inskip's interim report.

get almost unlimited money somehow, and that either nothing very horrid will happen, or if it did, it would not happen for a great many years."[40]

He went on to express the view that the results of such deficit spending would be felt sooner and more profoundly "than these cheerful people suppose."

The Cabinet meeting at which the interim report was considered brought few surprises. Duff Cooper stated that, in view of the fact that the report did not support the Admiralty's request that the New Standard be funded, he would have to ask for authorization for some destroyers and submarines not in the authorized program. Swinton expressed extreme displeasure with the suggestion that the nation and the R.A.F. should rely wholly on defence, and reiterated the importance of strategic bombing. He warned that the Government could not accept anything less than the proposals for Scheme J without a reversal in policy. Hore-Belisha accepted the change in the Army's role without apparent demur, which is indicative of the state of his relations with his general staff. Even Anthony Eden accepted, although with reservations, the elimination of the Army's continental role. He urged that any such actions should be taken only after consulting France, something Chamberlain was not about to do. He also expressed concern that the report meant Britain would be relying too much on the defensive, and supported Swinton in his bid to have Scheme J accepted. In closing the meeting Chamberlain reiterated his belief in the importance of the nation's stability as a deterrent to a potential attacker, and "hoped that the Cabinet would take a very special note of this aspect of the Minister's paper."[41] In its conclusions the Cabinet authorized the elimination of the Army's continental role and its reorientation to imperial and home defence. The Cabinet agreed to put off any decision on the Admiralty's ongoing request for the New Standard, and to submit the question of whether the fighter program should be emphasized over the bomber program to Inskip and Swinton for consultation.

[40] T 161/855/48431/01/2, 12/17/37, Bridges memorandum concerning Inskip's interim report.
[41] CAB 23/90A, 49 (37) 1, 12/22/37.

What the interim report had failed to do, and what the Treasury had hoped the whole review of defence spending would accomplish, was to establish concrete budgetary limits within which each service would have to stay. This plan had not come to pass because the figures the services submitted to the Treasury and Inskip in October to base the review on were so distorted as to be useless.[42] As a result the services were asked to submit new, more complete figures to the Treasury, along with their estimates, in January, and Inskip's final report was delayed until February. The importance the Treasury attached to Inskip arriving at a ration for each of the services is conveyed in this passage written by Bridges.

> I feel quite certain that unless we get a system of rationing, the Defence Programme will just go on growing, and we shall be powerless to stop them. I feel more strongly on this point than I did when we started this hare nine months ago.[43]

In the period before Inskip completed the final report, the Treasury was faced with the task of once more battling the services over the estimates for the 1938–39 budget. All the interim report had enabled them to do was "to survey the field and get some footholds for the next tug-of-war."[44]

As the Treasury waited for the services to submit their estimates, the tone of determination in its memoranda began to shade into one of desperation. Bridges set forth the rationale by which the Treasury planned to guide itself in the upcoming struggle.

> The main point is, while avoiding any prejudice to the review, to give the Departments a figure big enough to carry on the existing programme, but not big enough to enable them to embark on any new fancies.

> This is a fore-taste of rationing, and the Defence Departments cannot be expected to like it. But I am convinced that, unless

---

[42] T 161/855/48431/01/1, 11/23/37, Inskip note to the Treasury.
[43] T 161/855/48431/01/2, 12/17/37, Bridges memorandum on Inskip's interim report.
[44] Ibid.

we act firmly on these lines, at what I judge to be a very critical moment, we shall find that defence expenditure will overwhelm us. Every day I find further evidence of the continuing growth of defence expenditure on objects we cannot possibly prove to be unnecessary. There is only one answer: to say that there is a given sum of money available and no more. Now is the time to act.[45]

Both Fisher and Hopkins concurred.

I think this line of argument is sound, [wrote Fisher]. We all want to be safe, and this includes economic stability and staying power as well as bayonets. My military colleagues do not realise this and think of limitless "money" and not in terms of value for money.[46]

When the defence estimates came they were as high as the Treasury had feared they would be. There were grounds for suspecting that they had been padded to improve the respective services' positions in the bargaining, but even taking that probability into consideration, the Treasury found them discouraging.

The cause of the Treasury's feeling of helplessness in the face of increasing defence costs is to be found in the phrase in Bridges' note above, "objects we cannot possibly prove unnecessary." What is being alluded to is the Treasury's increasing inability to challenge the programs of the services on technical grounds, because of the escalating technological expertise required to evaluate those programs, and the services' monopoly of that expertise. They were becoming resigned to "the fact that criticism on merits by the Treasury of individual defence proposals always fails irrespective of its justification."[47] Warren Fisher asserted that:

Because individuals may wear a uniform and are called, e.g. the Air Staff, it does not follow that they are infallible guides. In regard to major principles, experienced laymen have a role to play.[48]

[45] T 161/821/34610/38, 1/10/38, Bridges on the defence estimates.
[46] T 161/821/34610/38, 1/13/38, Fisher handwritten note on Hopkins paper discussing Bridges memorandum.
[47] T 171/340, 1/14/38, Bridges memorandum on the 1938 estimates.
[48] T 161/855/48431/01/2, 12/18/37, Fisher note on Inskip's interim report.

Despite Fisher's assertion, the Treasury found it increasingly difficult to challenge on specific technical grounds a program submitted by a defence ministry, and have that challenge upheld by the Cabinet. Because of this, they resorted to their own sphere of expertise, finance, to negate such requests. Bridges used this metaphor to explain:

> in present conditions, the sort of criticism which the Treasury can bring to bear cannot be expected to bring down more than a few stray birds; the main covey will always get over—unless and until Departments are told that they can have £X millions and no more.[49]

Such rationing was calculated to force the services to evaluate their own programs and make the decisions on their importance within the context of the Treasury's limits. This approach had the virtue of compelling the departments with the requisite expertise to do the necessary evaluating. The services, however, resisted on the grounds that such financial rationing could force them to eliminate programs vital to the security of the nation. While Inskip struggled to arrive at a formula that would limit spending without jeopardizing defence, the Treasury concluded that it would have to impose a ration of its own to cope with the 1938 estimates.

Because the Air Ministry's programs were expanding most rapidly in both cost and technological refinement, they had become the major source of frustration to the Treasury. This frustration is reflected in the following comment by Hopkins concerning the 1938 estimates.

> The Air Ministry blindly pursue the policy of forever following (but doomed never to catch) Germany, plane for plane, type for type, and bomb for bomb. Enemies before now have been beaten by superiority of insight rather than by equality of numbers—attained or unattained. . . . there is no end to the money an unimaginative Air Ministry could attempt to get. Strategically there seems to me much to criticise in the way they propose to spend their allotment; financially the only question is what can be found for them.[50]

[49] T 161/855/48431/01/3, 2/14/38, Bridges on Inskip's final report.
[50] T 161/855/48431/01/3, 2/11/38, Hopkins memorandum on Inskip's final report.

In the same vein Fisher complained that, "The Air Staff have enslaved themselves to the parrot-cry of 'parity.' "[51] The Air Ministry had its own problems, and coping with the Treasury was but one of them. The Air Ministry was continually at odds with the aircraft producers as production fell behind, and this was exacerbated by the knowledge that the production gap between England and Germany was steadily widening. The Air Force estimates for the coming year were based on a modification of Scheme J called Scheme K, which was calculated to keep the production gap as narrow as possible. But even with Scheme K Swinton was beginning to assert that Britain could maintain her air security only through the implementation of controls.

The Admiralty too tried to extract as much as it could from the Treasury at this time. Hopkins wrote:

> The Admiralty case for introducing at once so vast a programme on top of two other programmes which were themselves very great will no doubt be justified by the present state of the world. That argument leaves me cold. . . . To those who agree as I do both that the aims of the Admiralty are far above the true case and that they are beyond our means, it is obvious that a halt ought to be called.[52]

In the course of February the first Lord of the Admiralty, Duff Cooper, fought long and hard with both Simon and Inskip for approval of the Admiralty program. He argued, as Hopkins predicted he would, that the world situation was critical and that failure to adopt the Navy scheme would be a sign that Britain was letting up in her resolve to be strong. The Treasury was equally adamant on the importance of not committing the nation to the construction of armaments she could not afford, and eventually the Treasury prevailed.

Even the Army, stripped as it had been of its continental role, presented figures that were larger than expected. The causes were the cost of creating the capacity to produce anti-aircraft guns, and the surprisingly high cost of the regular army's new colonial role.

---

[51] T 161/855/48431/01/2, 12/18/37, Fisher memorandum on Inskip's interim report.

[52] T 175/97, 1/19/38, Hopkins to Fisher on the Admiralty estimates.

Hopkins explained that the latter had been expected to be quite reasonable until, "the Foreign Office intervened with a terrifying picture of an . . . arduous African campaign which once more swelled the bill."[53] The Treasury had little trouble forcing reductions on the Army, as Hore-Belisha still lacked the support of a coherent staff.

In January the Treasury decided to demand cuts in the estimates of the services, not on technical grounds, but on the grounds that the authorization of spending beyond a certain amount would jeopardize the outcome of Inskip's review. Hopkins, Bridges, and Fisher decided to cut the Navy estimates by £6 million, those of the Army by £4 million, and those of the Air Force by £2 million. They calculated that this would permit the services to continue with authorized programs without enabling them to embark on new ones before Inskip was able to present his proposals.

INSKIP'S FINAL REPORT

On February 8, 1938, Thomas Inskip presented his final Report on Defence Expenditure in Future Years. Based on the figures submitted by the services in January, which had been revised to incorporate the Cabinet decisions taken as a result of his interim report, the report was, unlike the interim report, almost entirely the product of his own hand. The problem he confronted was monumental. Whereas the Treasury had fixed £1,500 million as the maximum cost the nation could possibly afford to pay for rearmament over the next five years, the total cost of the programs the services felt were necessary to defend the nation came to £1,926 million without the Navy's New Standard, and £1,968 million with it. His task was to bring the two figures into line without jeopardizing either the stability or the security of the nation.

He began by pointing out that there were two obvious courses that were possible in light of the incongruence of the figures.

> The first involves heavily increased taxation, and a straining of our economic system, leading to another crisis, or a long and painful period of bad trade.

[53] T 161/855/48431/01/3, 2/11/38, Hopkins memorandum on Inskip's final report.

If this course were pursued, "it would appear that we must envisage war measures of compulsion on industry and labour, measures not only most difficult politically, but threatening the maintenance of that stability which it is an essential defence interest to preserve."

The second course, in so far as it might be interpreted as a decision to restrict the defence programmes, might react upon the prospects of negotiations, and might, therefore, be fraught with the danger of war.

He found neither alternative acceptable for the reasons stated. His reservation about the second course, however, marks a subtle but significant shift from his position in the interim report. While he accepted the Treasury view that negotiations should be undertaken with potential enemies in an endeavor to find a way in which their grievances and wants could be peaceably ameliorated, thus eliminating possible causes of conflict and making rearmament unnecessary, he added the caveat that unless this was done from a position of strength it would be nothing less than a temptation to war. This new emphasis on the importance of being and appearing strong in the conduct of a policy of appeasement is the element in Inskip's final report that differentiates it from the earlier one, and is the key to the proposals he sets forth.

The most important proposal in the report was that all programs that would bring about an increase in the production of armaments in 1938 and 1939 should be fully funded, and that defence spending in those two years should be as great as necessary to insure maximum production during that time. He recommended that the total allocated for defence spending over the five-year period be increased to £1,570 million, and that £80 million more be allocated for air raid precautions. The allocation of the funds between the services would be decided in meetings between the Chancellor of the Exchequer, the Minister for the Co-ordination of Defence, and the three service ministers.

Inskip's proposal that the bulk of the spending over the five year period be concentrated in the coming two years was based on the hope that Britain could negotiate the appeasement of her potential enemies from a position of strength during that time, eliminating the need for spending on the longer-range projects. If Britain failed in her foreign policy, additional spending would become necessary, and of course the cost of the five-year program would greatly exceed the £1,570 Inskip projected. The virtue of the plan was that, in the event of the failure of appeasement, the nation would not be left unarmed.

It is apparent that Inskip drew up this compromise in response to admonitions by Hankey that the pursuit of appeasement, while slowing down rearmament, posed a threat to the nation's safety. In mid-January Hankey warned Inskip:

> Among the members, therefore, of what may be described as the "Great General Staff" there is a feeling of *malaise*, not yet formulated in writing, that our foreign policy, combined with our unpreparedness for war compared with some other countries, is putting us in a very dangerous position.

He also forcefully put forth his opinion:

> that we are at a parting of the ways. Either we must change our foreign policy or increase the *tempo* of our rearmament by dropping the principle that trade and industry are not to be interfered with. . . . Unless the number of our potential enemies can be reduced *at once* I submit that our rearmament policy must be changed. . . . A decision to press forward our armaments without regard to trade considerations might prove a decisive factor in deterring war or, if it came, in averting disaster.[54]

Although Inskip could not accept Hankey's conclusions, they made him aware of the importance of not risking the well-being of the country on the hope that appeasement could be achieved from a position of weakness. Inskip's plan gave appeasement two years to succeed without diminishing the existing rate of arms production. If that policy failed to achieve its ends during that period of time, Inskip felt that the nation would have to pursue the course advocated by Hankey. Politically he had no choice but to formulate a plan that would fall within the Treasury's prescribed financial limits but, as the man responsible for national defence, he

[54] CAB 63/53, MO (38) 2, 1/14/38, a note from Hankey to Inskip on foreign policy and defence.

was unwilling to put forth a plan that would put Britain at risk. The statement with which he concluded the report clearly reveals his hopes and fears.

> The plain fact which cannot be obscured is that it is beyond the resources of this country to make proper provision in peace for the defence of the British Empire against three different Powers in three different theaters of war. If the test should come, I have confidence in the power and inherent capacity of our race to prevail in the end. But the burden in peacetime of taking the steps which we are advised—I believe rightly—are prudent and indeed necessary in present circumstances, is too great for us. I therefore repeat with fresh emphasis the opinion which I have already expressed as to the importance of reducing the scale of our commitments and the number of our potential enemies.[55]

On February 16 the Cabinet discussed Inskip's paper at length. John Simon stated the Treasury view

> That the expenditure of £1,650 millions on defence not only placed a terrible strain on the national finances, but could not be increased without financial disorganisation to an extent that would weaken the resistance of the country.

In supporting Inskip's paper he mentioned neither the Treasury's unhappiness about the pattern of the proposed spending, nor its concern about the condition that the Treasury's financial limits could only be maintained if the policy of appeasement succeeded. Hopkins' view, expressed in a Treasury paper analyzing Inskip's report, was that the condition making the financial limitation dependent on the success of appeasement was "far too grave a risk to run, if there is any choice in view."[56] In the Cabinet Simon simply stressed the importance of adopting Inskip's program without any additions.

Of the service ministers, Duff Cooper and Hore-Belisha expressed the view that their departments could live with the proposals of the report. Swinton, however, was distinctly dissatisfied. He stated that it was the belief of the Air Staff that the proposed

[55] CAB 24/274, CP 24 (38), 2/8/38, Inskip's final Report on Defence Expenditure in Future Years.
[56] T 161/855/48431/01/3, 2/11/38, Hopkins paper on Inskip's final report.

financial limitations would not allow adequate preparations for the protection of Britain. He also pointed out that, due to the amount of time it took to get airplanes into production, it was necessary to place contracts as much as two years in advance of production, and questioned the provision that defence contracts not be placed beyond 1939. Although this point was not pursued at this time, it later became a real bone of contention.

Anthony Eden, who was to resign as Secretary of State for Foreign Affairs four days later, expressed the Foreign Office's view that the world situation would probably become worse, not better, in the course of 1938, and that the chance of reducing the number of Britain's potential enemies within the time suggested in Inskip's report was remote. It was an assessment with which the Prime Minister did not agree.

Chamberlain perceptively observed that the report was "to some extent an evasion and a postponement of the decision. . . . [Chamberlain] did not conceal that in the end this postponement might involve later on a decision of the very gravest character: nothing less, indeed, than a disastrous deterioration in the financial and credit situation." He said that he felt some doubt whether such a postponement was justified, and "had only brought himself to it by his hope for some improvement on the international side. We had been working for this for some time, and he thought there was a possible chance for an alleviation of the situation."[57] In Chamberlain's eyes the dangers inherent in financial collapse justified taking limited risks with national defence that he hoped could be ameliorated by a conciliatory foreign policy.

The Cabinet accepted the report, and empowered Inskip and Simon to consult with the service ministers on the allocation of funds. The services were to draw up plans, the total cost of which would not exceed the £1,570 million recommended in the report, and the final decisions were to be made after they were presented.

THE 1938 BUDGET

While Inskip's report was being prepared and discussed, the Treasury was grappling with its annual problem of finding sufficient revenue to pay for the defence programs that had been au-

[57] CAB 23/92, 5 (38) 9, 2/16/38.

thorized. On the positive side revenue for 1937–38 was £20 million more than they had estimated it would be as the result of the continued economic boom. In addition, the services had been unable to spend £34 million of the £80 million that had been borrowed for defence in 1937 because the defence industries lacked sufficient plant to absorb that much capital.[58] The grim reality, however, was that there was a deficit of £112 million between the estimated revenue and defence expenditure for 1938–39 despite the £54 million windfall from the previous year. This raised the inevitable question of whether that deficit should be met wholly by borrowing, or whether only £80 million should be borrowed and the balance raised out of new taxation. Richard Hopkins felt that taxation should be increased to meet part of the deficit. Besides his concern that the level of taxation be sufficiently high to cover the cost of maintaining the defence establishment when the rearmament program was completed, he had doubts about the Government's credit. He had hoped to be able to borrow, using long- and medium-term bonds, but he feared that

> if people see the total programme rising and no corresponding effort in taxation they will scent a coming glut of Government paper and a coming fall in security values (of which there was a warning last autumn) and be correspondingly slow to lend. Then we shall be forced back on financing ourselves temporarily on Treasury Bills. . . . We cannot permit a large and persistent addition of Treasury Bills for defence borrowings without misgivings as to the atmosphere it will create and the tendencies towards falling Government credit and a rising cost of living it will set in motion.[59]

As the time approached for a final decision, John Simon wrote a note setting forth the more political considerations involved in borrowing as opposed to taxation. He listed five considerations that favored a tax increase. The first two coincided with Hopkins' concerns stated above. In addition he pointed out that, as 1939 was expected to be an election year and the defence budget would

[58] T 171/340, 1/4/38, budget out-turn for 1937 from Inland Revenue.
[59] T 161/855/48431/01/3, 2/11/38, Hopkins paper on Inskip's final report.

increase again, making a tax increase inevitable in one of the two years anyway, such an increase should be undertaken immediately because an election year is "the worst year to choose for beginning increased taxation." Such an increase would also show the dictators that "we are determined to see things through." More importantly from the Treasury point of view:

> Further taxation imposed now, might drive into the heads of the spending Departments the idea that bigger and better ships and guns, lavishly estimated and impetuously demanded, are not an unmixed blessing, and that we are really at our wits end to pay for them. On the other hand, after all our Treasury protests, a comparatively "popular" Budget may only encourage the spending Departments to believe that there is plenty of money left.

The considerations weighing against increasing taxation were not quite as compelling. Simon pointed out that to ask for a tax increase after announcing the surpluses from the previous year would raise some eyebrows at budget time. Then too it would be perfectly feasible to get through 1938 without a tax increase, although not through 1939. Of greater import was the fact that the business community had expressed its opposition to a further tax increase and its preference for borrowing to meet the costs of the rearmament program. If the business community were to put up resistance to a tax increase, the dictators might believe that Britain could no longer maintain the pace in the arms race. There was also the danger that a tax increase might touch off a business recession, which was the last thing the Government wanted.[60] In the event it was decided that the tax increase was necessary, the bulk of which came from the income tax that was raised to five shillings, six pence in the pound, the highest level in the inter-war period.

By March the Chamberlain Government had completed the plans by which it hoped to avoid the perils posed to Britain's security and her solvency by the deteriorating international situation. Based on the Treasury's firm belief that a continuance of the exist-

[60] T 171/340, 3/16/38, Simon note on whether to borrow or tax.

ing rate of rearmament would destroy Britain's economy and consequently her ability to defend herself, the Government's program sought to limit the nation's expenditure on armaments while pursuing a foreign policy that would diminish her need for them. The policy by which defence spending was to be limited was known as rationing. The policy by which the nation's enemies were to be conciliated was known as appeasement. Both were rooted in the Government's desire to maintain the economic and social status quo. The fall of Austria on March 13, the Anschluss, was the first of a series of tests that would determine the success or failure of those policies.

CHAPTER VI

The Anschluss and the Czech Crisis: The Acid Test of Rationing and Appeasement

THE ANSCHLUSS

THE Anschluss, Germany's incorporation of Austria into the Reich by force, came at an awkward time for the Chamberlain Government. Just when the Cabinet had accepted the necessity of rationing the expenditure of the services, and tacitly approved the conduct of a conciliatory foreign policy as a consequence, Germany's violent act brought the efficacy of both courses into question. In the field of diplomacy the Anschluss was a rude blow to Chamberlain, who had been personally courting the friendship of the Italian leadership with an eye to winning their support in opposing this union of Austria and Germany. Chamberlain's efforts to gain Italy's favor had cost him the services of Anthony Eden a month earlier, precipitating a crisis within the Government. With Italy's acceptance of Germany's move, he had little to show for the risks he had taken. His first attempt to employ appeasement as an instrument of foreign policy had not borne fruit. The strategic implications of the Anschluss were even more disturbing than the act itself, and were, in time, to have much wider repercussions. The absorption of Austria put Germany in a position to dominate militarily a Czechoslovakia plagued by problems with the large German minority within her borders—problems, it might be added, that the German government played a considerable role in precipitating. The Cabinet's primary concern, from the time of its emergency meeting while the German invasion of Austria was still taking place, was with how Czechoslovakia could be saved from Austria's fate.[1]

It was a concern that jibed rather poorly with its recent decision that defence spending should be limited. That decision had been taken as the result of two full years of relatively good international

[1] CAB 23/92, 12 (38) 12, 3/12/38.

behavior by the Germans, which had created a climate in which it was possible to believe that they could be appealed to by reason tempered by conciliation, as well as by force. The violent aspect of the Anschluss caused some doubts as to the validity of this assumption, which inevitably raised doubts about the wisdom of placing financial restrictions on the rearmament program. For the Treasury, which was in the process of negotiating the individual service rations on which it placed so much importance, those doubts came at an especially unfortunate juncture. The Treasury feared that Germany's action would result in the product of their effort to regain control over expenditure being snatched from their grasp at the last moment.

THE SERVICES BID FOR FUNDS

Their fears were not without grounds. As surely as the night follows the day, the services moved to reintroduce programs the Treasury had hoped had been heard of for the last time. At the Cabinet meetings of March 12 and 14, which were convened to consider what action the Government should take in light of the new international situation, each of the services proposed programs that would strengthen Britain's military posture. The March 12 meeting was begun with the reading of a letter from Duff Cooper, who was confined to bed by illness, in which he stated that nothing less than Cabinet approval of the commencement of work on the New Standard would convince the world of Britain's resolve to deal resolutely with the dictators. This rather predictable response from the First Lord of the Admiralty was dismissed lightly by Chamberlain, who allowed that additional air power rather than additional sea power would be of more avail in influencing the course of events in landlocked Central Europe. Hore-Belisha then proposed that the Cabinet authorize additional expenditure on the construction of plant for the production of anti-aircraft guns and equipment in light of the serious deficiencies in the nation's anti-aircraft preparations.[2] Both Inskip and

[2] At a meeting of the C.I.D. four days later, Hore-Belisha stated that, due to the lack of productive capacity, only 7 of 352 3.7-inch anti-aircraft guns that had been ordered early the previous year had yet been delivered. CAB 2/7, 313th meeting, 3/17/38.

Swinton supported Hore-Belisha in his request, agreeing with his assessment of the seriousness of the situation. It was left to Swinton to introduce the proposal that was to raise the two issues the Treasury felt threatened the very economic foundations of the nation, and to lock the Government in controversy. The program he suggested was a modification of the once rejected Scheme K, designed to increase aircraft production by one third within three months. The problem was that the program would require both considerable additional expenditure and a massive increase in the number of skilled workers in the aircraft industry. To acquire these workers fast enough to have the desired impact on production would mean compelling firms with remunerative civilian contracts to release their skilled labor to the aircraft industry. Simon pointed out that the cost of the program would drive defence spending beyond the maximum approved by the Cabinet only a month earlier and would increase the danger "that we might knock our finances to pieces prematurely." Both the Secretary of State for Labour, Ernest Brown, and Inskip expressed the view that compulsion of labor and industry was not to be taken up lightly, as it would cause problems with both of those groups that could have serious consequences. Chamberlain agreed that it was desirable to increase aircraft production, but stressed that any acceleration would have to be nondisruptive, as interference with trade would be counter-productive. He suggested seeking methods other than compulsion to acquire skilled labor. Only Hore-Belisha, in an impassioned plea for decisive action, supported Swinton's contention that compulsion was necessary if the German threat was to be met.[3] While no decisions were reached at these two meetings, the issues that were to occupy the Cabinet's attention in the coming months were broached.

Although the Anschluss gave the services the opportunity to reintroduce programs that had previously been refused them, their success in having them approved was only slightly greater than before. In the course of the final discussions on the Army's budget for 1938, Hore-Belisha questioned the emphasis that Inskip's final report placed on the Army's role in home defence, and complained that the requirements for expanding the nation's anti-

[3] CAB 23/92, 13 (38) 3, 3/14/38.

aircraft defences would force the Army to take £37 million from other areas of its already shrunken budget to pay for them. The result would be a Field Force unequipped to carry out even its new Imperial role. Pointed out as an example of the Field Force's unpreparedness was the fact that the artillery it was using was a 1905 model whose range was only 6,000–9,000 yards as compared to that of modern equipment that had a range of 12,000 to 15,000 yards.[4] The protests were to little avail, however, as the C.I.D. upheld Inskip's recommendation concerning the Army's role, required the Army to comply with the Treasury's request that it cut its budget estimate of £347 million by £70 million, and commanded it to adhere to the mandated distribution of the remaining funds as well. This £70 million reduction was taken from three areas: the equipment of the T. A., the personnel of the colonial forces, and surprisingly, anti-aircraft equipment. The last item did not reflect a decision to ease up on air raid defence, but was dictated by the slow rate of output of anti-aircraft guns.[5] The Cabinet's authorization of these cuts indicated that the Government was determined not to alter Inskip's recommendations concerning the role of the Army, or Chamberlain's policy of not committing ground forces to the Continent implicit in it, despite the change in the European situation. While it was agreed that a powerful British and French military presence would be most helpful to the Czechs,[6] the decision was taken to eliminate any possibility of effective British ground intervention on the Continent. This indicates the Government's resolve to continue to take risks with its military strength in the hope that those risks could be ameliorated by a conciliatory foreign policy. Faced with the choice so clearly set forth by Hankey of taking risks with either finance or defence, the Government, at the insistence of the Prime Minister and the Treasury, chose the latter.

DUFF COOPER ATTACKS RATIONING

The most clearly stated and compelling attack on the principle of rationing, and the acceptance of risks to national defence that underlay it came from the Admiralty in the course of its efforts to

[4] CAB 2/7, 313th meeting, 3/17/38.
[5] CAB 24/276, CP 99 (38), 4/22/38, Inskip paper on War Office expenditure.
[6] CAB 23/93, 15 (38) 1, 3/22/38.

have the New Standard implemented. The Admiralty position was articulated by Duff Cooper. In a draft of a paper, written on defence expenditure in the future and its effects on the Admiralty's ability to carry out national policy, intended for submission to the Cabinet, Duff Cooper wrote the following:

> the more I reflect upon it the more convinced I become that *the system of rationing the Service Departments is impossible to defend*. I believe that it has never been adopted before. The very word "rationing" denotes a situation of extreme scarcity, which does not appear to have arisen. The first duty of a Government is to ensure adequate defences of the country. What these adequate defences are is certainly more easily ascertainable than the country's financial resources. The danger of underrating the former seems to me greater than the danger of overrating the latter, since the one may lead to defeat in war and complete destruction, whereas the other can only lead to severe embarrassment, heavy taxation, lowering of the standard of living and reduction of the social services. It seems to me that in this matter we should decide first upon our needs, for after all the Empire depends upon its defences, and that we should then enquire as to our means of meeting them. If it is found quite impossible to meet them, then it might be necessary to alter either the whole of our social system, or the whole of our foreign and imperial policy.[7] [The italics indicate a passage underlined by Chamberlain to whom the draft was submitted.]

Duff Cooper's final point about the need to alter the whole of the nation's foreign and imperial policy if the nation's resources should prove insufficient was an indirect reference to the Admiralty contention that, "so long as our ultimate naval strength remains undecided in principle, it is difficult to prepare war plans, to consider strategic dispositions, or to organise the required scale of recruiting and training. Such uncertainty prevents the Admiralty being able to guarantee the Navy's ability to implement our foreign policy."[8] The essence of the Admiralty's case was the questioning of the Treasury's contention that the funding of the New Standard would be an unbearable burden on the nation's re-

[7] PREM 1/346, 4/28/38, draft of CP 104 (38).
[8] PREM 1/346, 5/27/38, letter from Duff Cooper to Simon.

sources, and that if that contention were to be accepted immediate changes should be made to bring the nation's foreign policy into line with her military capabilities. More significantly, however, the Admiralty also raised the basic issue of whether financial and social stability should be a higher national priority than national security.

The Treasury was so profoundly disturbed by the Admiralty's questioning of the principle of rationing, and especially by the arguments set forth in Duff Cooper's draft paper, that it asked Simon to prevent the paper from being presented to the Cabinet. Hopkins wrote: "This document seems to me to be outrageous in its misstatement of facts and I do not see how the Chancellor could allow it to be circulated in its present form consistently with the responsibilities of his office."[9] He suggested that Simon should speak to Duff Cooper, "and persuade him of the error in his ways," rather than allow the paper to circulate even with a Treasury reply appended. The meeting apparently took place, but Duff Cooper was not persuaded. He continued to insist that the Cabinet be presented with the case, arguing that when it accepted Inskip's program in February, the Cabinet was not fully aware of the implications of its actions. Simon reluctantly agreed, writing to Chamberlain:

> It seems to me that we are reaching one of the supreme decisions and that unless it is decided that we cannot contemplate and provide in addition to everything else, a fleet the size of the New Standard Fleet, the whole of the Cabinet's decision of February 16th will soon be waste paper, and Inskip's effort at rationing collapses.

He continued:

> I am afraid that I have no very high opinion of their [the admirals who advised Duff Cooper] presumed view as to the possibility of the country bearing the immense burden they contemplate or of the consequences that would ensue if they have their way.[10]

Chamberlain agreed that the issue had to be presented to the Cabinet but in such a way as to insure that its decision would coincide

[9] PREM 1/346, Treasury response to Duff Cooper's draft of CP 104 (38).
[10] PREM 1/346, 6/24/38, Simon to Chamberlain.

with the Treasury view. It was decided, presumably by Chamberlain, that he, Simon, and Inskip should meet to discuss the case, and that Inskip, as the expert on defence matters, should write an evaluation of the Admiralty and Treasury cases for presentation to the Cabinet before the issue was discussed. Of course Chamberlain had previously ascertained that Inskip's paper would come down on the Treasury's side, and he assumed that the weight of Inskip's expertise, in combination with his own and the Chancellor of the Exchequer's political weight, would dispose the Cabinet to come to the proper conclusions. The climactic, if rigged, Cabinet meeting took place on July 20. Inskip led off by reiterating the conclusion of his paper that, though militarily desirable, the New Standard was financially impossible. Duff Cooper then proceeded to raise the points he made in the draft of C. P. 104 (38) that never was circulated to the Cabinet. He insisted that adequate provision for defence was the Government's first responsibility to the nation, and that rationing was nothing less than a limitation imposed on the nation's ability to defend itself. In conclusion he pointed out that the responsibility for the defensive weakness that resulted from rationing lay with the Cabinet, and expressed his doubt that it would be prepared to state publicly that it had decided to ignore the sea lords' advice on the Navy's requirements for defence for financial reasons. Simon's reply was that the Admiralty had been generously funded, and that if the New Standard were to be authorized there was every possibility that both the financial and social structures of the nation would be destroyed, leaving her incapable of defending herself no matter how many ships and guns she possessed. The outcome of the discussion was a foregone conclusion. It was decided to postpone any final decision on the New Standard. In a significant concession to the validity of Duff Cooper's argument, however, the Cabinet agreed that no hint should be given to the public that rearmament was being slowed in any way by financial considerations.[11]

CRITICISM OF THE AIR MINISTRY

While the Treasury succeeded in imposing the ration on the Army and the Navy, it had a more difficult time with the Air Force. At

[11] CAB 23/94, 33 (38) 8, 7/20/38.

the time of the Anschluss the situation of the R.A.F. was quite anomalous to begin with. Although the expansion program intended to create new productive capacity had been initiated two years earlier, for various reasons that capacity was only just beginning to produce. Consequently, from the point of view of monthly output and numbers of workers employed on production, Britain's figures compared miserably with those of a Germany producing at full capacity. Although Lord Weir pointed out in March that, "The performance for this month is not an accurate reflection of the total capacity of the industry,"[12] there was considerable criticism of the Air Ministry's management of the expansion program. Not surprisingly, the source of the criticism was the Society of British Aircraft Constructors, with whom the Air Ministry was in the final stages of their acrimonious two-year negotiation for an industry-wide contract. Because the two years of expansion had been initiated in haste and carried out under intense pressure for quick results, friction between the Air Ministry and the industry inevitably resulted from the delays that occurred. The industry blamed the Air Ministry for not providing information on future orders that was necessary if the industry was to prevent lapses between orders during which plant and labor would be unemployed. This was actually the result of the Cabinet's inability to decide on the role of the R.A.F., and the Treasury's unwillingness to authorize contracts more than twelve months in advance. They also cited the Air Ministry's insistence on changing the specifications of models already in production as an obvious cause of delays. These changes were necessary because, in order to shorten the length of time required to get plans into production from the drawing board, the Air Ministry had eliminated the building and testing of prototypes; hence, problems located in the first batches of planes produced had to be eliminated in the subsequent production runs.[13] For the industry's part, it was unwilling to stop work in the middle of contracts for planes that had become obsolete and retool for the production of more modern types at the request of the Air Ministry. This unwillingness stemmed from the fact that aircraft firms realized a higher rate of profit on

[12] Weir Papers 19/18, 2/38, Weir note defending the Air Ministry.
[13] Weir Papers 19/9, 4/20/38, copy of S.B.A.C. letter of complaint to Swinton.

models they were experienced in producing than when they began producing a new type. The industry's basic complaint was that the Air Ministry was arbitrary rather than cooperative, and that it refused even to consult with the industry to discuss ways in which cooperation could be improved. By late 1937 tension had built to the point that the S.B.A.C. was on the verge of publicly denouncing Swinton and the Air Ministry. This action was averted when the Government convinced the industry to appoint Charles Bruce-Gardiner as President of the S.B.A.C. to act as a liaison with the Government. While this appointment stemmed the immediate outburst, it did not result in the relief of the problems. Despite Bruce-Gardiner's appointment articles appeared in late January 1938 setting forth charges by some aircraft manufacturers that the Air Ministry was grossly inefficient and disruptive.[14] In the middle of February more criticism came from a different angle, but probably emanated from the same source. A parliamentary committee assigned to investigate the Air Ministry's administration of civil aviation issued a report in which it went out of its way to attack the Air Ministry's conduct of rearmament.[15] The relationship between the industry and the Air Ministry worsened to the point that Horace Wilson, acting for Chamberlain, began to deal directly with Bruce-Gardiner without Swinton's knowledge.[16] This arrangement gave the industry the opportunity to bring its grievances directly to the Prime Minister, seriously undercutting Swinton's position in dealing with them. At the heart of the industry's criticism of Swinton's and the Air Ministry's administration of the expansion program was its belief that its members were consulted too little and interfered with too much.

SWINTON'S MODEST PROPOSAL

As we have seen, Government interference with industry was one of the conditions that made the proposal for the rapid expansion of

[14] AIR 19/35, 2/1/38, press clippings of the critical articles. This file also contains some of the correspondence between Bruce-Gardiner and Swinton, but due to a lack of correspondence covering critical months, it is apparent that controversial material has been omitted from the file.

[15] CAB 27/643, CA 38 (3), 2/17/38.

[16] PREM 1/236, 3/15/38, a letter from Wilson to Chamberlain on the conversations he had with Bruce-Gardiner concerning S.B.A.C. complaints about the Air Ministry.

aircraft production, submitted to the Cabinet by Swinton in the wake of the Anschluss, so controversial. It should be remembered that Swinton, as early as 1936, had stated his belief that rapid expansion of aircraft production could only be achieved by means of compulsion. That belief was confirmed by his experience of the subsequent two years in trying to expand production without controls; and, the provisions of Scheme K that he first submitted to the Cabinet in January 1938 strongly hinted at the need to compel firms to permit the transference of their skilled labor to the aircraft industry. His suggestion at the March 14 Cabinet meeting, that 70,000 workers be transferred from the civil to the defence sector, therefore came as no surprise. The problem Swinton's proposal posed for the Cabinet was succinctly put by Lord Weir.

> Are we to place rapidity of rearmament and a further increase in defence strength in front of all other considerations, and enter on a field of central control, and conscription of capital, facilities and labour? Without this it is hard to see any scheme by which substantial savings in time and production rates in excess of existing estimates can be secured.[17]

Swinton's argument for compulsion of labor was that the capacity to increase the production of planes had at last been completed, and all that was needed was the labor to operate it on a double-shift basis. Neither the Prime Minister nor his industrial adviser, Horace Wilson, found the prospect of compulsion either politically or economically palatable, and had already taken steps to increase the amount of skilled labor available to the defence sector by voluntary means.[18]

In a speech to Parliament on March 24, announcing the Government's intention to expand armaments production as a result of the Anschluss, Chamberlain gave an indication of the approach

---

[17] Weir Papers 19/18, undated but probably about 3/13/38, notes for a meeting with Inskip and Swinton.

[18] It should be noted that, after the Cabinet ruled against Swinton's recommendation that controls be imposed to secure the skilled labor required for the expansion of aircraft production at the March 14 meeting, he accepted the decision and took an active part in the Government's efforts to gain the cooperation of organized labor. See, for example, SWIN 270/40, 3/22/38, minutes of the 118th meeting of the Air Council, pp. 29-32, and Weir Papers 19/1, 4/1/38, Swinton to Horace Wilson on the labor shortage in engineering.

the government intended to use to increase the supply of labor when he stated:

> Steps are already being taken to inform organised workers and organised employers of the nature of the demands which the accelerated plans will make upon their industries, and thus place them in a position to devise practical methods for meeting those demands by mutual arrangements, and with a minimum of Government interference.[19]

In short, the Government had decided that it would continue to use nothing more than cooperation in its efforts to accelerate armaments production, despite the advice of its own specialists at the Air Ministry that, unless compulsion were employed, no significant increase in production rates could be expected. One important change in existing Government policy was revealed in Chamberlain's statement. For the first time in its history, the National Government deigned to bring labor into consultation.[20] The Government required the cooperation of organized labor if it were to succeed in shifting large numbers of skilled workers into the defence industry without causing widespread labor unrest.

The decision that Horace Wilson should discuss with Walter Citrine, the General Secretary of the Trades Union Congress, the possibility of the Government's approaching the T.U.C. in order to gain its cooperation in expediting the execution of the rearmament program was made before the German invasion of Austria.[21] The discussion took place the following week. In the course of it Wilson stressed that what the Government desired was the general support of the T.U.C., and that it opposed any government interference in the relations between local unions and employers. What the Government actually wanted was the T.U.C. to do what it could to persuade local unions to cooperate with employers in the efforts that would be necessary to accelerate arms production. Wilson also expressly stated the Government's wish that the T.U.C. do this as quietly as possible, and make every effort to keep the Labour Party ignorant of it. He came away from the

---

[19] CAB 4/27, 1416-B, 3/24/38, extracts from the Prime Minister's statement in Parliament.

[20] See Chapter III, pp. 125-128.

[21] PREM 1/251, 3/12/38, note from Hankey to Horace Wilson.

meeting with Citrine's assurance that he would raise the question with the T.U.C., and with the impression that labor was really unaware of the seriousness of either the international or the defence situations.[22] Citrine was successful in persuading the leadership of the T.U.C. to consider the matter, and subsequent meetings were arranged with Chamberlain in which the whole situation was discussed. As a result of those meetings agreement was reached that the T.U.C. would do what it could to expedite the transfer of workers into the defence sector. By June the T.U.C. had come under fire from some unions, such as the Amalgamated Engineering Union, on the grounds that it was failing to represent adequately their members' interests to the Government in the disputes that arose with management over the changes being made in working conditions.[23] While the Government insisted on staying out of any local disputes, it did consider means of aiding "helpful elements" in the engineers' union to persuade their colleagues to cooperate.

While Wilson was trying to win the support of the T.U.C., he and Inskip were also consulting with the employers for the purpose of devising a plan to expedite the transfer of labor to the defence sector. The first man consulted was Sir Alexander Ramsey, the President of the Employers' Federation. He informed Inskip that, if labor cooperated, the employers could make arrangements for the shifting of skilled workers among themselves, but that labor's cooperation was vital. Wilson then spoke to Charles Bruce-Gardiner, whose S.B.A.C. stood to be the primary beneficiary of the program. The S.B.A.C. had favored the use of Government compulsion to secure the labor it required, but Wilson informed Bruce-Gardiner that the Government preferred "to get the best we can out of a joint voluntary effort by the firms and the Trade Unions."[24] At the end of March Bruce-Gardiner, Ramsey, Weir, and Wilson met to arrange the process by which employers would release workers to defence industries, especially

[22] PREM 1/251, 3/21/38, note from Wilson to Chamberlain on his meeting with Citrine.

[23] PREM 1/251, 6/38, various memoranda. For a detailed account of the engineers' complaints against the T.U.C. and the Government, see *The New Statesman*, "The Engineers and the Government," 5/14/38, pp. 820-821.

[24] PREM 1/236, 3/18/38, letter from Wilson to Bruce-Gardiner.

the aircraft industry. The actual transfers were to be worked out between the firms themselves. The workers affected were to be allowed some pay increases, although they obviously would not be as great as those they would have been able to command if the industries were bidding competitively for their services.[25] The lesser increases prevented the wage spiral the Government feared would lead to inflation, and kept the prices it had to pay for armaments at a lower level as well.

The Government's effort in trying to expand the amount of labor available for defence work without resorting to compulsion or costly outright competition between the civilian and defence sectors is a classic example of the doctrine of cooperation at work. While this cooperation served the Government and the aircraft industry well, its benefits for the workers, who received token pay increases and no guarantees of security analogous to those that the aircraft industry extracted in return for its cooperation, are less obvious. The benefits for the employers who agreed to release workers to the defence industries are also moot. In reply to the charge that the Employer's Federation and the T.U.C. did less than they should have on behalf of their memberships, it can only be said that if they had refused to cooperate as they did the Government would have been forced to use compulsory powers to achieve the same ends. Regardless of the motivation of the two groups in cooperating with the Government, the S.B.A.C.'s rather self-serving behavior throughout the rearmament program contrasted poorly with that cooperation.

It should be noted that the transfer of skilled labor was not the only method used to deal with the labor shortage. Under the agreement made between the Government and the T.U.C., unions permitted the training of greater numbers of workers in certain highly skilled jobs, a practice known as dilution, which they would normally have struck to resist. They also permitted the breaking down of some skilled jobs into a number of less skilled operations, a process known as de-skilling, which permitted the industry to increase its output without increasing its demand for skilled workers.

[25] PREM 1/251, 3/29-31/38, memorandum on the meetings of Wilson, Weir, Ramsey, and Bruce-Gardiner.

Probably the most important method of expanding the output of the aircraft industry was the extensive use of subcontracting. In late March Inskip broached to Swinton the possibility of "bringing work to labour" through the extension of subcontracting procedures. Swinton replied at that time that this plan would not meet the immediate needs because the necessary plant was already in place and required sufficient labor to work it around the clock. At the same time the aircraft industry was also resistant to subcontracting for a more prosaic reason. This passage from a memorandum by R. H. Phillips, the Director of Aeronautical Production at the Air Ministry, to the Air Ministry Supply Officer explains the industry's reluctance:

> they have no intention of putting out to be made by others, any part which they could possibly make by themselves and so make a higher profit. . . . Their profits on sub-contracted parts are strictly controlled and are much smaller, as a general rule, than on parts they make themselves. It probably pays them in the long run to make as many parts as possible of their own for their aircraft.

Phillips then goes on to explain that the optimally efficient use of resources would dictate that aircraft plants be used wholly as assembly points, for they alone are equipped for that process, while the construction of all the parts would be subcontracted to outside firms. This arrangement would greatly expand the number of planes that could be assembled, reduce the amount that had to be spent on tools and gauges (as it would bring the work to the existing equipment), and save the Government a good deal of the money it was currently spending to equip complete aircraft factories fully. Phillips concludes that:

> The attitude of the industry to this subject is, I consider, of major importance not only politically but to the carrying out of the programme. . . . While it is not the policy of the Department to interfere with industry more than can be helped . . . unless something is done in connection with this question of sub-contracting, we shall not attain anything like the aircraft we hope for under the accelerated programme.[26]

[26] AIR 2/2711, 5/25/38, D.A.P. to A.M.S.O. on subcontracting.

In the course of the summer of 1938 pressure was brought to bear on the industry by the Government to persuade it to subcontract out more of its parts, resulting in a major increase in the number of planes Britain was able to produce. The plan proved to be the most productive innovation the Government made in these years, enabling it to squeeze sufficient output from the existing productive structure and to avoid resorting to controls.

THE DEBATE OVER R.A.F. EXPENDITURE

How the Government would dispose of Swinton's proposal to use compulsion to acquire the labor necessary to carry out the acceleration of the Air Ministry's production program was really never in any doubt, but the question of dimensions and cost became the center of a bitter and closely contested battle between the Air Ministry and the Treasury. After the March 14 Cabinet meeting Inskip and Swinton sat down together to work out a program that would enable Britain to keep pace with Germany in aircraft production, with an emphasis on the rapid acceleration of production in the short term. The plan that emerged, a modification of Scheme K calling for fewer bombers and more fighters, was referred to as Scheme L. The Treasury's two major objections to this scheme were that it called for the placement of contracts more than twelve months in advance and, more importantly, that it called for an increase of £60 million over the £507 million that had been decided on as the Air Ministry's share of the overall five-year defence ration. The Treasury's basic contention remained that the nation could not afford it. In a paper presented to the Cabinet on Scheme L Simon wrote:

> Recent events in Europe, serious as they are, have done nothing to increase the financial resources of this country, to diminish the vital importance of those resources as an essential element in our defensive strength, or to change the force of the proposition approved by the Cabinet in February last that £1,650 millions (of which £1,570 millions is for the three Defence Services) was the maximum which could be regarded as available over the five years 1937–41.[27]

[27] CAB 24/276, CP 87 (38), 4/4/38, Simon on Scheme L.

211

The Treasury continued to question the need for the bomber program, and proposed an alternative program for the R.A.F. that would keep its expenditure within the ration. That program would have provided more air strength than Scheme F, but considerably less than Scheme L, which was cited by the Air Ministry as the absolute minimum Britain could undertake if she wished to remain within hailing distance of Germany in aircraft production. The Treasury's contention was that the Air Ministry should strive to create a small but efficient force from existing resources and draw up paper plans for implementation in the event that the international situation should take a turn for the worse. "Readiness as soon as possible" became the Treasury's criterion for what the Air Ministry should aim at.

It fell to Inskip to weigh the arguments on both sides and present a paper to the Cabinet containing his conclusions. In his evaluation Inskip agreed with the Treasury on the issue of cost, and with the Air Ministry that Scheme L was the minimum program consistent with the requirements of national security. Only in the annex to the paper did Inskip give a hint as to which consideration he believed should be given priority.[28] He waited until the Cabinet meeting of April 6, at which the issue was to be settled, to announce his support for the Air Ministry view. Having been satisfied that Scheme L could be carried out without resort to the compulsion of labor, he expressed his belief that the risks to the nation involved in not maintaining sufficient air strength to contest Germany were greater than those involved in increasing defence spending by £60 million. Chamberlain followed Inskip, and asked Swinton what the effect of a Cabinet rejection of Scheme L would be. Swinton was unequivocal in his reply. Scheme L was a minimum, dictated not by what the Air Ministry considered would provide Britain with safety from a strategic point of view, but by political considerations of what was possible without controls. "Politically anything less was a confession of complete and permanent inferiority to Germany." He stressed the importance of working to a long-term plan if the nation's resources were to be efficiently used, and emphasized the need for immediate authori-

[28] CAB 24/276, CP 86 (38), 4/1/38, Inskip on the Treasury and Air Ministry positions.

zation of such a plan, mentioning that the Cabinet had been reviewing and rejecting Air Ministry plans for more than a year.

> We should be deceiving ourselves and others [he stated] if we pretended that we could speed up or act later if we postponed action now. He contended that now was the time: but in any case quick action was impossible later unless orders were given and action taken now.

In reply to the contention that Scheme L would create labor problems that would make its execution difficult, he reiterated that the capacity to produce the planes was in existence, and that it would be a dereliction of duty to fail to employ it fully. To the Treasury's suggestion that the nation could get along with a smaller, more efficient air force, he replied that efficiency was no substitute for effectiveness. In summing up, he declared that to reject the Air Ministry's program on the grounds of economy would be a false saving because Scheme L represented both the nation's last hope for peace and, failing that, her only hope in war.

Swinton's was a compelling performance, and Simon's rebuttal was weak, familiar, and unconvincing.

> First, the essential principle was to preserve the financial strength of the country. . . . These figures for the cost of Scheme "L" could not be reached unless we turned ourselves into a different kind of nation. Germany, for example, had got rid of her war debt; had not such good social services as we; controlled wages and could enforce loans. . . . We could not do these things. Consequently there was a strict limit to our expenditure.[29]

The Cabinet was faced with an impossible choice: to support Swinton and Inskip, whose case was overwhelming, and turn on the two political leaders of the Government, or to once more support their political leaders and quite possibly doom the nation. In reality the Cabinet never makes such choices. It was agreed that Chamberlain, Simon, Swinton, and Inskip should reach an agreement among themselves and submit it to the Cabinet for approval.

[29] CAB 23/92, 18 (38) 8, 4/6/38.

Powerful as Swinton's presentation of his case was, the crucial element that prevented the Cabinet from supporting the Treasury as it had always done before was Inskip's support. He had come to be seen by his colleagues as the impartial expert on defence issues. Throughout 1937 he had been convinced of the validity of the Treasury's contention that there was a readily discernible limit to the amount the nation could afford to spend on defence, and he had played the central role in creating the mechanism by which service expenditure would be kept within that limit. In the period between the issuance of his interim and final Reports on Defence Expenditure in Future Years, his view of the situation began to change, probably as the result of Hankey's warnings. He began to place more importance on the nation's defensive strength, and became less impressed by the Treasury's cataclysmic predictions about the effects of exceeding the mandated limit on defence spending. This shift is reflected in the subtle but important differences between his interim and final reports. The Anschluss probably served to reinforce his conviction that the nation simply could not afford to take any greater risks on defence. While he still opposed the use of compulsion, which he knew would have an immediate effect on the economic and social stability of the nation, he had come to the realization that the Luftwaffe posed a more profound and immediate threat to Britain than did the increases in spending necessary to counter it. His support of Scheme L, in direct opposition to the Treasury, marked the completion of his shift from the view that financial stability should be the Government's first priority.

Because his colleagues in the Cabinet had come to respect the scope and depth of Inskip's grasp of the defence situation, a respect encouraged by the Treasury when he had wholly supported its point of view, he had become in effect the Cabinet's arbiter on questions concerning defence. As Duff Cooper put it after Inskip's paper had sealed the fate of the New Standard:

> I am sure that it was never intended when a Minister for Co-Ordination of Defence was created that our future naval policy should be based on the opinion of a very well meaning and terribly over-worked barrister. But unfortunately we are all so over-worked that most of our colleagues, who clearly have not

the time themselves to consider great questions which do not concern their departments, are inclined to think that because Tom is a very good chap and a very fair-minded man, his opinion must be the last word and that they had better let it go at that.[30]

That is an accurate assessment of Inskip's influence on defence questions. When he threw that influence behind Swinton, it was sufficient to offset the opinions of the two most politically powerful figures in the Cabinet, the Chancellor of the Exchequer and the Prime Minister. The significance of Inskip's emergence as an independent force in the determination of defence policy should not be overlooked. It altered the political balance within the Cabinet, providing an effective counterweight to the Treasury's heretofore unchallenged authority on matters pertaining to defence spending.

Although there are no minutes of the meetings that took place among Simon, Swinton, Inskip, and Chamberlain on the subject of the new Air Ministry acceleration program, it is possible to piece together from their results the compromises that were made. The Treasury had entered the meetings determined that, "The immediate objective must be to complete and make effective what has already been taken in hand."[31] It was opposed to adding a definite figure to the amount that it had already specified as the absolute limit for defence spending over the five-year period, and strenuously opposed the placement of contracts for new projects. The Air Ministry was convinced that the nation's production of planes had to be increased if the German threat were to be met, that to accomplish this compulsion of labor would probably be necessary, and that the aircraft industry had to be given contracts for production beyond the next twelve months if it was to be expected to go about such an expansion of production.

As we have seen, the Government decided not to resort to compulsion to acquire the labor necessary for the acceleration of aircraft production. The Air Ministry was also forced to forego the placement of any contracts beyond the next two years, but Swinton was upheld in his primary contention that more planes were

[30] Cambridge University Library, Templewood Papers, X 1-5, 7/19/38, a letter from Duff Cooper to Samuel Hoare.
[31] T 161/923/40760/03/1, 4/7/38, Simon note on the Air Ministry program.

vital to the national defence. The Treasury agreed that "what really mattered was to get maximum production of aircraft from British manufacturers during the next two years. Whatever they could produce during that time we ought to take."[32] Under this plan the Air Ministry was empowered to order 12,000 planes for completion by March 1940, rather than the 7,500 previously authorized, and the Treasury accepted the fact that its cost would exceed the limit it had hoped to impose on Air Ministry spending.[33] Inskip's hand in this compromise is clear. It is in keeping with the strategy set forth in his final report that the nation should rearm as fully as possible over the coming two years in the hope of being able to reach an accommodation with her enemies from a position of relative strength during that time. Both the decisions to spend what was necessary to purchase everything the aircraft industry had the capacity to produce within the next two years and to limit all contracts to projects to be completed within that time span comply with Inskip's strategy. While Inskip formulated the compromise, its acceptance by Chamberlain was necessary if it were to be adopted. That he did accept it makes it clear that, like Inskip, he opposed compulsion, but was no longer convinced that the dangers of exceeding the Treasury's defence ration were greater than those the ration posed to the nation's ability to defend herself. This is not to say that either he or Inskip rejected the Treasury view that defence spending must be limited if the economy was to remain viable, but simply that they realized that the Treasury limit was not absolute, and that the risks of somewhat exceeding it were outweighed by the resulting benefit to the nation's security. Without Inskip's intervention it is doubtful that Chamberlain would have come to this point of view and certain that the acceleration of aircraft production would not have been authorized. While the doctrine of rationing was still very much in force, Inskip had caused it to be tempered to accommodate the harshest realities of the times. Inskip's achievement was a small one on the face of it, but it is doubtful whether Britain would have survived the Battle of Britain without it.

[32] T 161/923/40760/03/1, 4/25/38, Simon to Fisher on the compromise on the Air Ministry program.

[33] CAB 23/93, 21 (38) 6, 4/27/38.

Inskip was unquestionably aided in his task of persuading Chamberlain to accede to the compromise by the increasing criticism in both Parliament and the press of the Government's rearmament program. The leading voice in that criticism was Churchill's, as it had been consistently since 1934. For the first time, however, he was joined by the parliamentary Labour Party. This change was important because it made it difficult for the Government to dismiss the new wave of criticism of the effectiveness and adequacy of its rearmament program as merely another example of Churchillian alarmism. Although their doubts about the state of British rearmament were motivated by different concerns—Churchill's as the result of a nationalist concern about the survival of the British state in the face of an attack by a foreign power, and Labour's as the result of an ideological concern about the survival of working-class socialism in the face of fascism—and although their positions on virtually every other issue of the day were diametrically opposed, the convergence of their views on this subject caused the public to question seriously for the first time the assurances the Government had been giving about the progress of rearmament.

The groundwork for Labour's turnabout had been laid in 1936 when some of the leaders of organized labor, most importantly Ernest Bevin and Walter Citrine, along with a segment of the parliamentary Labour Party led by Hugh Dalton came to the conclusions that fascist Germany posed a serious threat not only to Britain but to the working-class movement in Britain, that the League of Nations would be ineffective in opposing any German resort to force, and therefore that Britain had to rearm sufficiently to counter the fascist threat. They were opposed on this issue by the pacifist element in the parliamentary Labour Party which included Clement Attlee, its leader, as well as by the extreme left wing of the movement, led by Stafford Cripps. As noted in Chapter II, the parliamentary Labour Party had voted against the Government's defence estimates in 1936 and 1937 on the grounds that they were excessive and not formulated with any reference to collective security or the League of Nations. The party's criticism in the

Commons concentrated on the expense of the Government's rearmament programs, and the scope they offered for profiteering by industry. In October 1937, Dalton and Bevin worked to have a resolution passed at the party conference at Bournemouth that stated in part that, "we must confront potential aggressors with an emphatic superiority of armed strength."[34] Their success resulted in a substantial modification in the party's previous position, and cleared the way for Labour members of Parliament to challenge the adequacy of the Government's rearmament policy in the Commons. It was not until after the Anschluss and the commencement of the Czech crisis in the early summer of 1938, however, that the party actually pressed this line of criticism. When it did, in concert with Churchill, the press took it up, and for the first time the Government came under real pressure from the public concerning its conduct of the rearmament program.[35]

Criticism reached its peak in May when the Air Ministry's failures in the execution of the expansion program became the subject of a sharp debate in the Commons. While Churchill had been challenging the Government on the subject for more than a year, Labour's entry on his side, in combination with new charges by aircraft manufacturers concerning Air Ministry interference with production, injected new heat into the exchanges, and made it impossible for the Government to slough off the question as it had been able to when Churchill had been pressing it alone. The aircraft manufacturers' charges, which, as we have seen, stemmed from old grievances, were initiated when the S.B.A.C. learned that the Air Ministry intended to send Lord Weir to Canada to look into the possibility of purchasing aircraft there to supplement those being produced in Britain. They were angered by this because they felt that the Air Ministry had not made sufficient effort to realize all the production it could from the capacity of Britain's aircraft industry, and they argued that they could produce all the

[34] *Parliamentary Debates* (House of Commons), Vol. 336, 5/25/38, col. 1235.

[35] For a review of the development of Labour's position on foreign policy and rearmament during the period see John Naylor's *Labour's International Policy: The Labour Party in the 1930's*. Also of use are Michael R. Gordon, *Conflict and Consensus in Labour's Foreign Policy, 1914-1965* (Stanford: Stanford University Press, 1969), pp. 62-82, and G.D.H. Cole, *A History of the Labour Party from 1914* (London: Routledge and Kegan Paul, 1948), pp. 320-365.

aircraft that Britain needed if the Air Ministry would stop interfering with their production schedules.[36] At an emergency meeting of the S.B.A.C. on April 20 it was proposed that the industry make a public attack on Swinton, a proposal that Bruce-Gardiner was able to prevent from being implemented at the last moment by promising to bring the industry's grievances directly to the Government. Although this promise forestalled a statement by the S.B.A.C., it did not prevent individual manufacturers from making their complaints public.[37] It was these charges that fueled the debate about the Air Ministry in which Labour played such an active role.

On May 12, 1938, the public outcry against the shortcomings of the Air Ministry's expansion program reached its peak. In the course of debate in Parliament, a coalition of dissident Tory backbenchers led by Churchill joined the Labour opposition to wring from the Government an admission that the expansion program was considerably behind schedule, and that Britain was vulnerable as a consequence. Soon thereafter Labour moved for a public inquiry into the management of the Air Ministry that was supported by twenty-seven Conservatives. While the Government had no trouble defeating the motion, it was evident to Chamberlain that dissatisfaction with the Government's rearmament policy within the ranks of his own party was spreading, and that steps would have to be taken if a political crisis was to be avoided.

SWINTON SACKED

Wasting no time, Chamberlain requested Swinton's resignation on the thirteenth and received it the following day. In the brief public letter that accompanied his resignation Swinton explained that he was stepping aside to allow the Prime Minister to replace him with someone with a seat in the House of Commons who would be able to respond directly to the mounting criticism that the Air Ministry was being subjected to there. Although this letter implied that Chamberlain had asked Swinton to step down in re-

[36] T 161/923/40760/03/1, 4/25/38, Simon note to Fisher in which he refers to what Horace Wilson had relayed to Chamberlain from the S.B.A.C.
[37] PREM 1/236, 4/21/38, notes from Wilson to Chamberlain on the aircraft industry's unhappiness with Swinton.

sponse to political pressure stemming from the criticism of the expansion program, it is clear that there were more significant underlying considerations as well. After all, Chamberlain had fully supported Swinton in the face of similar criticism only three months earlier.

In the intervening period, however, circumstances had caused Chamberlain to reconsider Swinton's role in the Cabinet. Among those circumstances was the further deterioration of the relationship between the Treasury and the Air Ministry as a result of the latter's successful effort to persuade the Cabinet to override the Treasury's attempts to impose the ration on its expenditures. The Treasury retaliated with criticism of the Air Ministry's efficiency that was very similar in content to the S.B.A.C.'s complaints. In one acid note Fisher wrote:

> For some years now we have had from the Air Ministry soothing-syrup and incompetence in equal measure. For the first time in centuries our country is (and must continue to be) at the mercy of a foreign power.[38]

There is a certain irony in the Treasury's taking the side of the aircraft industry against the Air Ministry when in fact the Treasury was the root cause of the grievances for which the industry held the Air Ministry responsible. Ironic or not, the Treasury's antipathy toward the Air Ministry, and especially that of Simon toward Swinton, who regularly demolished the Chancellor of the Exchequer in debate in the Cabinet, was a definite factor in Chamberlain's decision to request Swinton's resignation.

Probably the most important factor was Swinton's increasingly insistent advocacy of the need for compulsion to meet the German threat. On this subject he clashed directly with Chamberlain's own strongly held views. Chamberlain felt that such a course would not only mean political disaster because it would antagonize both labor and business, but he was certain that it would lead swiftly to economic chaos as well. He believed that more remained to be accomplished by means of cooperation, and Swinton's continued insistence on compulsion, after Chamberlain had made this view clear, could not have pleased him. This advocacy

[38] PREM 1/252, 4/2/38, note from Fisher to Hankey.

had not been enough to warrant his dismissal in the past, however (it will be remembered that he had suggested the use of compulsion as early as 1936), and it is doubtful that if he had remained isolated in it that Chamberlain would have regarded it as more than an annoyance. The problem was that, while Swinton was compulsion's most insistent and outspoken advocate within the Cabinet, there were others like Duff Cooper and Hore-Belisha who were beginning to give signs of agreeing with him that the time was approaching for its use. What made his departure certain, however, was Inskip's swing to his support on the question of Scheme L. This alliance of the two most expert and compelling voices on issues of defence threatened Chamberlain's leadership and control of the Cabinet. While he agreed in this case with the argument that Britain required more air power, and knew that Inskip shared his views on the subject of compulsion, Chamberlain could not dismiss the possibility that at some future date Inskip might shift to Swinton's view on compulsion as he had just done on the subject of the ration. In such an event the combination of Inskip's honesty and sincerity with Swinton's compelling persuasiveness could have resulted in a struggle for the support of the Cabinet and possibly for the leadership of the Government as well. The prospect was not one that Chamberlain could be expected to tolerate. As Swinton had the personal qualities to lead such a political insurrection if presented with the opportunity, and since he was vulnerable at the time, Chamberlain took the opportunity to replace him with a talented and trusted colleague, Kingsley Wood. Although Chamberlain offered Swinton his choice of any other position in the Cabinet, in the hope of forestalling further political problems, Swinton declined. For all the tenacity with which he fought for his views in the Cabinet, Swinton refused to make his resignation a political issue, preferring instead to go quietly, a loyal party man to the end. Only Lord Weir, Swinton's close friend and adviser at the Air Ministry, publicly objected to Swinton's sacking, resigning his position as adviser to the Government in protest.[39]

The circumstances surrounding both men's resignations have tended to obscure their contributions to Britain's rearmament ef-

[39] Roskill, *Hankey*, Vol. III, pp. 355-356.

fort. In their three years at the Air Ministry they conceived and supervised the conversion of the British aircraft industry to large-scale production involving the most advanced aircraft technology, often in the face of staunch resistance from the industry itself. When they resigned, their efforts were just on the verge of fruition. During the ensuing two years the aircraft that won the battle of Britain were built in the factories that Swinton and Weir had brought into being.[40]

KINGSLEY WOOD AND THE TREASURY

Although Swinton's departure may have kindled some rejoicing at the Treasury, it was short lived. Kingsley Wood, the new Secretary of State for Air and former Minister for Health, was one of the few men in the Cabinet whose opinions Chamberlain listened to. He very quickly took control of the Air Ministry, and within a month was besieging the Treasury with requests for funds. In his very first communication with the Treasury, five days after his appointment, he requested authorization to order 13,800 planes for delivery by March 1940, 1,800 more than the Cabinet had authorized in the recent compromise, on the assumption that the industry would fall 15 percent short of promised deliveries. He also quickly initiated talks with Lord Nuffield, the automobile manufacturer who had earlier refused to enter into the shadow scheme as the result of a conflict with Swinton, concerning the possibility that he would now agree to undertake to build airplanes for the Air Ministry. As the two approached an understanding on the subject, Chamberlain entered into the discussions, and agreement was soon reached that Nuffield would construct a large plant to produce fighters. The Air Ministry was to give him an extraordinarily

---

[40] Of the important members of the Chamberlain Government there is less known about Swinton than about any of the others save Inskip. He wrote a rather unrevealing memoir, *I Remember* (London: Hutchinson, 1948) and (in collaboration with James Margach) a short book of reminiscences of some of the famous men he worked with entitled *Sixty Years of Power* (London: Hutchinson, 1966). The Air Ministry and Cabinet files contain few of his papers, and his private papers at Churchill College in Cambridge contain very little on his tenure at the Air Ministry either. Consequently it is difficult to determine the considerations that underlay some of the controversial positions he took while serving as Secretary of State for Air.

large initial order of 1,000 planes, a free hand in choosing the location for his factory, and the promise of no interference in his plans for production. That the new Secretary of State never consulted the Treasury in the course of these negotiations, presenting it with a *fait accompli* at their conclusion did little to endear him to it. The Treasury's initial response to Wood's zeal was a patient note from Simon in which he gently suggested that they should get together to reach, "a clearer understanding than exists at present between the Treasury and the Air Ministry as to what is contemplated and what is consequential upon arrangements now being made."[41] The two apparently failed to reach this understanding because a month later the Chancellor of the Exchequer felt constrained to lecture Wood at a meeting of ministers that

> it was imperative that normal Treasury financial control and responsibility in regard to the numerous and growing activities of the Air Ministry should be re-established without delay in order that relations between the Treasury and the Air Ministry in regard to financial control of the Treasury over expenditure by the Air Ministry should be assimilated to the arrangements in force in regard to other spending Departments.[42]

Kingsley Wood continued to pressure Simon for funds. In late July after Wood had come to him with his latest request for £10 million to expand the war potential of the aircraft industry, a request he accompanied with "a light hearted warning that it will be £40 million before he has finished," Simon wrote Chamberlain a despairing note. He concluded it by saying:

> I have now done all I should. . . . It seems to me that we must boldly scotch the idea of a pledge of parity as meaning the German figures however they grow. It is impossible: it is unnecessary: and it is quite inconsistent with rationing, which K. W. fights shy of.[43]

[41] T 161/923/40760/03/1, 5/24/34, note from Simon to Wood.
[42] CAB 24/277, CP 148 (38), 6/24/38, minutes of a conference of ministers on the Air Ministry's mission to Canada.
[43] PREM 1/236, 7/29/38, Simon to Chamberlain on Kingsley Wood's continued pressure for funds.

From the Treasury point of view the new man at the Air Ministry was no improvement on the old one. In a rather lucid analysis of the prospects for holding the Air Ministry in any kind of financial check, Bridges pointed out that for 1939 and 1940 the Air Ministry was scheduled to spend £125 and £175 million respectively, leaving it an average of £92.5 million to spend in the last two years of the program if it were to stay within the Treasury's ration. Wrote Bridges:

> Such a sudden drop is almost inconceivable. It is true that there is, no doubt, a very sporting chance that, even with the help of the aircraft industry in the Gorgeous West, the Air Ministry will not spend £175 millions next year going all out. But against this the Secretary of State for Air . . . will not necessarily rest content with the scheme now put forward. He may well press for still further expansion . . . it is obvious that Sir Kingsley Wood does not regard his present plans are anything more than the best he can do at the moment.[44]

Just as the Treasury "seemed at last to be getting a firm hold upon a ration . . . it slipped its leash once more."[45] The prospects for the 1939 budget were grim indeed, as were those for keeping defence spending within the £1,650 million five-year limit. Still the Treasury clung to the ration. Uncertain as it was, the Treasury saw no other prospect for maintaining the Government's and the nation's solvency. For the Treasury the coming year was to be a time of profound disappointment.

MUNICH AND REARMAMENT

In the course of the summer the attention of the Cabinet turned from the battle between the Treasury and the services to the Czech crisis, which became the first major test of the Government's strategies of rationing and appeasement. Since the German annexation of Austria in March, it had been expected that Czechoslovakia would become the next object of Hitler's expansionist intentions.[46] The wave of protest from the German population in the

[44] T 161/923/40760/03/2, 7/18/38, Bridges on R.A.F. spending after 1940.
[45] T 171/341, 6/24/38, Hopkins on the budget for 1939-40.
[46] There are many detailed accounts of the diplomatic maneuverings leading to

Sudetenland against their mistreatment by the Czech government that erupted six weeks after the Anschluss signalled the beginning of this campaign. Provoked and led by German agents, the Sudeten Germans presented the Czechs with a list of their demands. When the German government began to threaten to intervene militarily on behalf of the Sudeten Germans if those demands were not satisfactorily met, the affair was transformed into a European crisis. The reason for the transformation was the series of alliances that France had arranged to keep Germany in check. Under those alliances France was bound to assist Czechoslovakia if she were subjected to German aggression, and Britain was similarly bound to support France in any conflict she became involved in with Germany. Consequently Hitler's talk of military intervention turned what had appeared to be only a regional dispute into a situation that threatened to drag all of Europe into war.

While Britain and France put up a bold front, warning Germany that they fully intended to stand by their treaty obligations, both had many reservations about confronting Germany militarily, and pressured the Czechs to reach a quick settlement with their dissident German minority. Despite that pressure and the efforts of a British mediator, Walter Runciman, no such settlement was reached as the Sudeten Germans at Hitler's command turned down every Czech proposal. By early September talks had broken down. Tension mounted as Hitler's venemous speech at the great Nuremberg Nazi party rally on September 12 touched off incidents of violence between Germans and Czechs in the Sudetenland. When the Czech government sent in troops to restore order, German intervention seemed imminent. At this moment, when Europe seemed to be on the brink of war, Neville Chamberlain initiated his personal efforts to save the peace. Hitler accepted his offer to fly to Germany to arrange a peaceful settlement of the Czech issue, and on September 15 they met at Berchtesgaden for the first of their three encounters. While the results of that meeting

---

Munich. John Wheeler-Bennett's *Munich, Prologue to Tragedy* (London: Macmillan, 1948) provides an accurate and readable description of the crisis, while Keith Robbins' *Munich* (London: Cassell, 1968) gives a deeper, more rounded account. Keith Middlemas' *Diplomacy of Illusion*, which is based on sources that have only recently become available, integrates the details of the diplomatic aspect of the crisis into the strategic and political context in which it took place.

gave Chamberlain reason to believe that an agreement was possible, Hitler's intransigence at their second meeting, at Godesberg on September 22, seemed to destroy all hope. War appeared to be inevitable, and the people of Europe began to prepare for the worst. In one final effort to reach a settlement Chamberlain flew to Munich on September 29. There he bargained away the last guarantees of Czechoslovakian independence for what he hoped would be an enduring peace. A relieved and grateful Europe hailed the Munich agreements, and bestowed their praise on the Prime Minister who made them possible.

While Chamberlain has been condemned for pursuing his policy of appeasement in reaching the Munich agreements, at the time he was dealing with the crisis he did not believe he had any viable alternatives. The evaluation of the military situation presented to him by the C.O.S. detailed in the starkest terms the extent of Britain's impotence. In it the C.O.S. pointed out that if Germany attacked Czechoslovakia with the bulk of her land and air forces she would still be sufficiently strong to turn back any French attacks in the West. As a result of the Government's earlier decision not to equip the Army for activity on the Continent, the British were unable to offer France any ground assistance in such an endeavor, as they required the full complement of their troops to man the defences at home. The C.O.S. also considered that it would be unwise to send the R.A.F. on missions against Germany, largely because the German striking capacity was twice that of the British and French combined. (The British estimated that the combined British and French air forces could deliver 300 tons of bombs per day as opposed to Germany's estimated capacity of 600 tons per day.) In addition the Air Ministry was unwilling to risk having its bomber force, which was completely lacking reserves, decimated in an early onslaught. The effectiveness of such an onslaught would have been limited in any case because, "The majority of our present striking force consists of bombers whose range . . . enables them to penetrate only a short distance into Germany."[47] The C.O.S. concluded that "no pressure we and France could bring to bear, by sea, land or air could stop

[47] CAB 16/183A, DP (P) 34, 10/24/38, part of the C.O.S.'s reassessment of the nation's defensive preparedness in light of the Munich crisis.

Germany from overrunning Bohemia, and inflicting a decisive defeat on Cz[echoslovakia],"[48] that any war that was engaged upon would be a long one, and that in the event that Italy or Japan entered it on Germany's side it would, "place a dangerous strain on the resources of the Empire."[49] Although these conclusions could not have come as a surprise to Chamberlain, for they represented the logical consequence of the rearmament policy he had done so much to shape, they no doubt spurred him in his efforts to find a peaceful solution to the Czech crisis.

It should be remembered, however, that the policy by which he chose to pursue that solution, appeasement, was not devised in the heat of the moment by politicians anxious to save their nation from having to fight a war they lacked the courage to face. Rather it was conceived, as we have seen, in 1937 as a necessary corollary to the Government's decision to ration defence expenditure. Both the defence ration and appeasement were the result of the Government's carefully calculated assessment of the economic, social, political, and strategic realities that Britain faced, and not of any weakness in the character of her leaders. While there is much to criticize in the assumptions that underlay that assessment, assumptions as to the source of Britain's strength and the resilience of both her people and her economy, they are criticisms that speak to questions of political and economic philosophy rather than to those of courage or cowardice.

Munich marked the beginning of the end of the Government's rearmament policy. While the bankruptcy of that policy became increasingly evident, the considerations that had shaped it obstructed efforts to recast it in the face of the increasingly menacing threat of war. For the Government 1939 was to be a year of disillusionment on every front.

[48] Churchill College Library, Inskip Diary 1, 9/14/38, summary of the C.O.S. view of Britain's prospects in a war with Germany over the Czech issue. The C.O.S. paper to which Inskip refers, C.O.S. 765, was not published for the Cabinet until after the crisis. It then appeared as CAB 16/183A, DP (P) 32, 10/4/38.

[49] CAB 16/183A, DP (P) 32, 10/4/38.

CHAPTER VII

From Munich to War: The Unraveling

PART I: THE AFTERMATH OF MUNICH

The International Situation

DESPITE the military weaknesses that were exposed by the Czech crisis, the conclusion of the Munich agreements could have been justifiably regarded by Chamberlain as a vindication of the rearmament policy he had initiated. He had succeeded through appeasement in bringing about an outcome he lacked the power to determine through strength, and had taken a step toward the conciliation of Britain's major enemy. All of this was in keeping with the imperative stated by the C.O.S. and reiterated in Inskip's final report that it was essential to reduce the number of the nation's potential enemies. Chamberlain saw Munich as a first step toward this end, and he was determined to follow it up. He wrote in his diary on November 6:

> In the past, I have often felt a sense of helpless exasperation at the way things have been allowed to drift in foreign affairs, but now I am in a position to keep them on the move, and while I am P.M. I don't mean to go to sleep.[1]

He felt that "Rome at the moment is the end of the Axis on which it is easiest to make an impression,"[2] and arranged talks with Mussolini in January. Although the talks were convivial, they bore no fruit, and Britain was no closer to winning Italian friendship than she had been before.

Shortly after Chamberlain's return from Rome, Halifax set before the Cabinet the Foreign Office's view of the behavior of Germany in the near future. Citing Hitler's frustration at being robbed of a glorious military victory by the Munich settlement, and the deterioration of the German economy, Halifax explained that Hitler was faced with the choice of slowing down German rearmament and raising taxes, the course the German indus-

[1] Feiling, *Chamberlain*, p. 389.
[2] Ibid.

228

trialists were pressuring him to pursue, or of staging some kind of "explosion," the object of which "would be to distract attention from the failure of his system to work in time of peace." Such an "explosion" of course would be in the form of a military adventure that would permit him to suppress the political elements that were trying to influence his conduct of policy, and to seize new supplies of raw materials. Halifax went on to say that, while it had previously been believed that such a move would be directed to the East, the latest intelligence was that he had shifted his focus to the West, and concluded by stating:

> All that can be said with practical certainty is that an "explosion" of Germany is likely to come in the comparatively near future, and that it is necessary for us to take immediate measures to guard against the possibility of its being directed against us.[3]

These were not welcome words to Chamberlain, who maintained his belief that Germany had no aggressive intentions and could be appeased. In the Cabinet he strongly discounted the Foreign Office's analysis of the German situation, stressing that, in his talks with the German ambassador, Von Dirksen, he had been assured that Hitler had the economic situation well under control.

While Chamberlain was attempting to improve Britain's relations with Italy and Germany, her closest ally, France, was pressing his Government to expand its recently renewed commitment to provide a ground force to aid her in any armed confrontation she might have on the Continent with those same two powers. The French General Staff contended that, because the British had bargained away thirty-six Czech divisions, divisions upon which the French had previously counted to help offset Germany's increasingly powerful ground forces, they were responsible for providing "an army of a size to do something to redress the balance between France and Germany."[4] The French considered this a matter of considerable urgency for, like the British Foreign Office, they believed a German attack was imminent. Within France there was increasing resentment toward the British based largely on the be-

[3] CAB 23/97, 2 (39) 1, 1/25/39.

[4] Inskip Diary 2, 1/11/39, summation of a report from the British military attaché in Paris on the French attitude concerning British aid.

lief that "Great Britain is very willing to fight her battles in Europe with French soldiers."[5] This French pressure reopened the question of the role of the British Army, and the old battles between the War Office and the Treasury were renewed, with the Foreign Office coming down heavily on the War Office's side.

The Reaction at Home

The Cabinet was not the only place that questions were raised about the wisdom of hewing to the same basic foreign and rearmament policies that had been pursued before Munich. The amorphous collection of Conservative backbenchers that had previously challenged the effectiveness of the rearmament program regarded Munich as a capitulation, and resolutely opposed the Government's efforts to carry on as before. This group, which included Churchill, Eden, and Duff Cooper (who had resigned from the Cabinet over the Munich agreements), led thirty Conservative MP's in abstaining on the motion to support the Government after the Munich debate. In the months that followed they spoke against and refused to support defence proposals they considered inadequate, often siding with the Opposition. They supported Labour's proposal to create a Ministry of Supply in November, and the following month five junior ministers from their ranks threatened to resign in protest against the inadequacy of the rearmament effort.[6]

Many of the dissidents came under heavy pressure from their local party organizations as a result of their opposition to the Government's policy, and the party whips tried to bring them into line with the threat of a snap election in which they would not receive the party's endorsement.[7] Such pressure was keenly felt, as the rebels believed that there was little public support for their views and that, if an election were called, few of their number would be returned.

Chamberlain was pressed by many Conservatives to take advantage of the wave of popularity, which they believed the party enjoyed immediately after Munich, to strengthen his hand and si-

[5] CAB 16/183A, DP (P) 47, 2/17/39, Foreign Office analysis of the French strategic situation.
[6] Middlemas, *Diplomacy of Illusion*, p. 452.
[7] Neville Thompson, *The Anti-Appeasers: Conservative Opposition to Appeasement in the 1930's* (Oxford: Clarendon Press, 1971), pp. 184-185.

lence his critics. Hoare put the argument for doing so in the following terms in a message to the Prime Minister:

> With the present House of Commons, and with Parliament nearing its end, you will not get a fair run for a policy of peace. Every set-back, and there may be many, will be exploited against you, and with the House of Commons already breaking into cliques, the situation may soon become intolerable.[8]

Chamberlain and the Conservative Party managers, however, were less than convinced that the electorate would be as enamored of the Government and appeasement after a hard political campaign as they were in early October when they had just been spared from war. Their caution was borne out by the results of the six by-elections held in October and November, in which voting swung against the Government by an average of 5.7 percent and two seats were lost. Public opinion polls taken at the time also indicated that support for the Government was not as strong as was widely supposed.[9] Whether political instinct or more substantial information warned Chamberlain against a campaign is difficult to say. Whatever the reason, he decided he could carry through his policies and deal with their critics most effectively with his existing Government.

Foreign Policy after Prague

Only after the German invasion of Czechoslovakia on March 15, 1939, was Chamberlain forced by the great outpouring of public sentiment to abandon appeasement.[10] Although both he and Simon initially took the position in the days following the fall of Prague that the event meant nothing, they were quickly persuaded by their colleagues that such a position was no longer politically viable, and that it was imperative for Britain to take a strong position against further German aggression.

---

[8] Middlemas, *Diplomacy of Illusion*, p. 417.

[9] Middlemas, *Diplomacy of Illusion*, p. 417, footnote. A Mass Observation poll taken in October showed 50 percent supported the Government's policy toward Czechoslovakia while 42 percent opposed it. Although the sample on which the poll was based was small, its results were consistent with those of a Gallup poll taken the following February in which 50 percent expressed their support for the Government, 42 percent stated opposition, and 8 percent did not know.

[10] Roger Eatwell points out that the public's attitude toward the Government and appeasement, which seemed to many historians to shift suddenly from appro-

In the weeks that followed the Government had ample opportunity to do precisely that as rumors flew from capital to capital concerning where Germany would strike next. On March 31, in response to Germany's increasingly threatening behavior toward Poland, Chamberlain announced to Parliament that, in the event of any German action that compromised Polish independence, "His Majesty's Government would feel themselves bound at once to lend the Polish government all support in their power."[11] This guarantee to Poland ran completely contrary to the Government's previous policy of increasing isolation and was all the more remarkable because Britain had no military force to support it, the whole orientation of the rearmament program being toward home defence. It can only be assumed that the guarantee was extended to demonstrate the Government's resolve to oppose German aggression not only to Hitler but, more importantly, to the British public.

During this same period Britain and France became involved in negotiations with the Russians concerning the possibility of forming an alliance against Germany. It is apparent from Britain's conduct of these negotiations, which lasted until mid-August when the Hitler-Stalin pact was hurriedly concluded, that the Government was engaging in them in order to silence those in its own party as well as those in the Opposition who were demanding such talks, rather than out of any desire for a treaty themselves. The Government also believed, mistakenly as it turned out, that their participation in these negotiations would preclude the possibility of the Russians making a treaty with the Germans. The losers in this charade were the French, who had an obvious interest in facing Germany with the threat of the Russian army on her eastern flank. The British forced them to choose between the possibility of a Russian alliance and the assurance of British aid, such as that was, in the event of a German attack by making it clear that, if they concluded a separate alliance with Russia and became

bation to opposition after the invasion of Czechoslovakia, had actually taken shape in the two months following Munich. The fall of Prague served to provide both a stimulus and focus for what had been the public's latent discontent with the Government's policy, transforming it into active opposition. "Munich, Public Opinion and Popular Front," *Journal of Contemporary History*, Vol. 6, no. 4 (1971), pp. 122-139.

[11] Mowat, *Britain*, p. 639.

involved in war with Germany as a result, Britain would not feel obliged to come to their assistance. This nice bit of diplomatic blackmail gave the British a free hand to conduct the negotiations to suit their own political purposes.[12]

As the English spring slowly turned to summer, Europe moved inexorably toward war. Although the imminence of that war was widely accepted—a question of when rather than whether— Britain's efforts to prepare for it were marked more by changes in form than by changes in substance, as the basic considerations that restrained British rearmament throughout the thirties operated until the actual declaration of war and beyond.

Rearmament after Munich

All of this lay very much in the future when Neville Chamberlain met with the Cabinet on October 3, after his triumphant return from Munich, to discuss the effect of the recent crisis on the rearmament program. At that meeting he expressed his belief that

> we were now in a more hopeful position, and that the contacts which had been established with the Dictator Powers opened up the possibility that we might be able to reach some agreement with them which would stop the armaments race. It was clear, however, that it would be madness for the country to stop rearming until we were convinced that other countries would act in the same way. For the time being, therefore, we should relax no particle of effort until our deficiencies had been made good. That, however, was not the same as to say that as a thanks offering for the present detente, we should at once embark on a great increase in our armaments programme.[13]

The deficiencies to which he alluded lay mainly in the area of home defence, and especially in Britain's anti-aircraft defences, which the recent crisis had shown to be woefully underequipped, undermanned, and unorganized. The services felt that the crisis had exposed unpreparedness in other spheres as well, and did not

---

[12] For a detailed account of the negotiations see Robert P. Shay, Jr., "The Death of a Hope: The Anglo-Soviet Negotiations of 1939" (Master's Essay, Columbia University, 1970). See also Robert Manne, "The British Decision for Alliance with Russia, May 1939," *Journal of Contemporary History*, Vol. 9, no. 3 (July 1974), who feels that the British began to negotiate in earnest in May.

[13] CAB 23/95, 48 (38) 5, 10/3/38.

hesitate to set forth their proposals to remedy them. It was soon decided, in keeping with the Prime Minister's opening statement, that, although no new programs should be embarked upon, already authorized programs should be accelerated in areas where the recent crisis had demonstrated Britain's preparation to be deficient. This decision of course did nothing to deter the services from presenting new programs that they felt were necessary in the guise of authorized programs that they wished to see accelerated, a ploy that fooled neither the Prime Minister nor the Treasury. In a note to Chamberlain Horace Wilson noted:

> Some large figures are in mind at the W. O. and at the Air Ministry and it would be well to get on record tomorrow [at the Cabinet meeting] that we must hesitate before departing too far from the aggregate sum approved, reluctantly, by the Cabinet earlier in the year.[14]

In the same note he also suggested that a cabinet-level committee be created to consider the services' accelerated plans before they were submitted to the Cabinet in order to screen out programs that would drive the defence budget beyond the previously agreed limits. Such a committee was formed, and it was made clear by Chamberlain and his advisers that the Government in no way felt that the recent crisis justified any increase in the ceiling on defence expenditure. That Chamberlain regarded Munich as a vindication of the policies of rationing and appeasement is readily apparent.

The only service that did not openly challenge Chamberlain's view was the Admiralty, whose new First Lord, Lord Stanhope, had previously been the Minister for Education. A firm believer in the existing rearmament policy, he was Duff Cooper's opposite in almost every sense. During the period between the time Stanhope became First Lord and the outbreak of war, the question of the New Standard was never raised in the Cabinet, although the Admiralty remained as vigorous as ever in its pursuit of funds. While the proposals they put forth for accelerating naval programs immediately following Munich were minor, reflecting perhaps Stanhope's unfamiliarity with his new position, the Admiralty es-

---

[14] PREM 1/252, 10/25/38, note from Wilson to Chamberlain.

timates presented in January for the coming fiscal year considerably exceeded the ration that had been allocated for the Navy the previous year. The Treasury reaction was predictably heated.

> We have always known that the Admiralty regarded the "new ration system" as something which should be got rid of at the earliest possible moment. It is apparently too much for the Board of Admiralty that an attempt should be made to relate our expenditure to our financial resources over a period of years. They are also apparently unaware—or if they are aware, choose to forget—that the Cabinet, both in February and in June, specifically upheld the system of rationing.[15]

Unlike his predecessor, Stanhope stayed within the normal channels in the bargaining that followed with the Treasury over the amount to be appropriated for the Navy. A firm and determined negotiator, he avoided dramatic appeals to the Cabinet, and raised none of the disturbing broader issues that it had been Duff Cooper's wont to introduce. The correspondence between him and Simon on the Navy estimates, with its cool and businesslike tone, contrasts markedly with the acrimonious exchanges between Duff Cooper and Simon. Uninclined and unwilling to rock the boat, Duff Stanhope was ideally suited to serve in the Chamberlain Cabinet.

The Army's Resurrection

While the Admiralty gained little from the tensions generated by the Munich crisis, the Army acquired a whole new lease on life. On September 13, 1938, as the crisis was building to its climax, the Cabinet restored the Army's continental role on the pre-1938 scale. It authorized the Army to prepare itself to be able to send two regular divisions and one mobile division to the Continent within twenty-one days, and two more regular divisions within forty days. The problem was that it appropriated no funds for the purpose, and left all the divisions in question equipped on the scale that had been authorized the previous February, that is, for Imperial defence. Thus when the time came for the services to introduce plans to remedy deficiencies, the War Office pointed out

[15] T 161/905/35171/39, 1/4/39, Treasury review of the Admiralty estimates.

that "present arrangements will not permit the Army to meet satisfactorily or safely the responsibilities it may be called upon to discharge in accordance with its approved role."[16] To bring the Field Force up to the standard of preparedness necessary to fight on the Continent, Hore-Belisha requested an appropriation of £30 million, and an additional appropriation of £11 million to equip a separate force for colonial duty.[17] He also requested funds to equip the Territorial Army as a reserve force. This proposal was put before the C.I.D. on December 15, and was strongly supported by Lord Halifax, the Secretary of State for Foreign Affairs, who indicated that if Britain did not take steps to create a Field Force that could effectively aid France against Germany, the French might well stand aside rather than bear the slaughter of facing Germany alone in a ground war.[18] Hore-Belisha and Halifax had hoped to impress the C.I.D. with the urgency of the matter and win its approval to present the issue to the Cabinet before Christmas, but neither Simon nor Inskip agreed that such haste was necessary. Both expressed the belief that it was preferable to permit the Treasury to study the proposal in the context of the rest of the defence estimates for 1939, and send it to the Cabinet along with the other estimates in January.

The Treasury took a great deal of time with the Army proposals, and did not present the War Office with its conclusions until late in January, allowing it very little time to revise its original plan before it was presented to the Cabinet. In the interim the French had increased their pressure for a substantial commitment of ground forces, and the concessions made to them in this matter meant that the War Office requirements were actually more expensive than those embodied in its original programs. The situation was presented to the Cabinet on February 2, and Hore-Belisha opened the discussion by stating that the War Office required £81 million more than it had been allotted in the previous year to meet effectively its recently revived continental responsibilities. Halifax again supported the War Office's request, reiterating the Foreign Office's belief that French cooperation with

[16] CAB 4/29, 1498-B, 12/13/38, Hore-Belisha memorandum on the Army's deficiencies.

[17] Hore-Belisha's relationship with his staff at this time, though still far from cordial, was at least functional.

[18] CAB 2/8, 34 (2), 12/15/38.

Britain would be very much affected by the size of the Field Force and the rapidity with which it could be dispatched. Chamberlain questioned the possibility that the French would not fight the Germans if they did not receive what they regarded as adequate ground support from Britain; and, he expressed his belief that the French would appreciate that, with the gigantic effort Britain was making with her Navy and Air Force, Britain could not afford to provide a large army as well. Simon then stressed the precariousness of the financial situation. He "entirely accepted the view that, in the present circumstances, there was no alternative but to use borrowing powers." He had to point out, however, that there were limits to the amount that could be borrowed, and he believed "it was impossible to escape the conclusion that we were advancing to a position in which the financial situation would get altogether out of hand." He therefore questioned "whether it was really necessary to adopt the proposals made by the Secretary of State for War, at any rate in their present form, and whether a substantial reduction could not be effected."[19] Although Hore-Belisha stressed the importance of an immediate decision, the Cabinet agreed that, in view of the lack of agreement on the part of the departments involved, the heads of those departments should meet with the Prime Minister and the Minister for the Co-ordination of Defence to work out a compromise for presentation to the Cabinet. The details of that meeting are not available, but it is clear from the recommendations that eventuated from it that the threat of the French not to fight was a telling factor. The Army was authorized to equip four regular divisions and two small mobile divisions for service on the Continent, and to equip four Territorial Army divisions to the point where they could be dispatched as reserves within six months of the departure of the regulars. Savings were made by deferring the equipment of the proposed two-division colonial force. While this compromise cost the Treasury £60 million more than it had planned to spend under the ration, it did not really assure that the British would be able to send an effective Field Force to France. The primary problem remained: Britain lacked the industrial capacity to produce the necessary guns and equipment for the Force. Until that capacity was created—by a process that would require a large scale finan-

19 CAB 27/97, 5 (39) 3, 2/2/39.

cial commitment and at least two years to complete—the Field Force would remain underequipped. Despite the expense of the compromise, the fact remained that it was really no more than a gesture to stiffen French resolve, a gesture the Government would have preferred not to have had to make, and one that did not really satisfy the French. They argued that nothing less than the implementation of military conscription would indicate genuine British commitment. At this time conscription was out of the question, however, and certain elements in the Government felt that they had been blackmailed into providing as much for the Field Force as they had.

Kingsley Wood's Frankenstein Monster

The Air Ministry also took advantage of the Munich crisis to press for a major expansion of its programs. Presenting figures that showed Britain lagging behind Germany by a margin of two to one in both first line strength and monthly output, and projecting a worsening of that situation in the course of the next year if existing programs were adhered to, Kingsley Wood proposed a massive new program designed to diminish the German advantage. Although it did not envisage the use of compulsion, the cost of the program was staggering, almost doubling the £123 million already authorized to be spent on the R.A.F. by 1940. Under the program orders would be placed for 3,700 fighters and 3,500 big bombers, most of which would be delivered after the April 1940 cutoff date the Air Ministry had agreed to in the spring compromise with the Treasury. Wood asked the Cabinet for authorization to place half of those orders immediately, stressing that they were vital if the aircraft industry were to be provided with the incentive to continue to expand its productive capacity.[20] Under the guise of an acceleration program, the Air Ministry was in fact trying to obtain authorization for a new bomber program entailing an expensive long-term spending commitment, as well as for a more immediate expansion of its fighter program.

The Air Ministry's intentions did not escape the notice of the Treasury, which objected both to the program's cost, and to the

[20] CAB 24/279, CP 218 (38), 10/25/38, Air Ministry case for strengthening the R.A.F.

fact that it would commit the Government to the expenditure of large amounts in the period after April 1940. The Treasury calculated that this new scheme would increase the Air Ministry's expenditure under the five-year expansion program from the £500 million limit under the ration to £850 million, which would in turn drive overall service spending from £1,650 million over the five years to £2,000 million.[21] In the Treasury's estimation this increase would be more than the nation could bear. In a paper to the Cabinet Simon explained:

> The Air Ministry programme is . . . so costly as to raise serious doubt whether it can be financed beyond 1939/40 without the gravest danger to the country's stability. The damage which I apprehend is not of the sort which can be got over by calling for "sacrifices"; it would consist of weakening our economic and financial strength as no increase in taxation could remedy. Excessive borrowing entails risk of higher costs, higher wages, and almost certainly higher interest rates so that the burden on the country even if tolerable at first becomes progressively worse. Moreover, it means substantially increased imports and substantially reduced exports. Our balance of payments— already a serious problem—will become more and more serious. In the end our monetary reserves (which have already been heavily depleted since the crisis by withdrawal of foreign capital from this country) might be still more rapidly exhausted and we should have lost the means of carrying on a long struggle altogether.

> I do not for a moment claim that purely financial considerations can have priority over urgent definite needs for material defence. The two things have to be considered together. The worst of all results would be to reach a position hereafter in which defence plans would be *openly seen* to have been frustrated by the financial and economic situation.[22]

In light of these possible problems the Treasury suggested that the Cabinet authorize and give top priority to the expansion of the

[21] T 161/923/40760/03/2, 10/29/38, Hopkins to Simon on the Treasury's analysis of the Air Ministry's proposals.

[22] CAB 24/280, CP 247 (38), Appendix #1, 11/3/38, Treasury recommendations concerning the Air Ministry program.

fighter program, while slowing the bomber program to the point where existing plant and labor were maintained, but no more. The Treasury also asked that it be given the right to review long-term orders and cancel them at any time if the financial situation so required.

On November 7th the Cabinet met to consider the Air Ministry's proposals and the Treasury's amendments. Kingsley Wood began by discussing the importance of Britain's air strength to the Prime Minister's international bargaining position, reminding the Cabinet members of the weaknesses that had been revealed in the course of the Munich crisis. He then turned to the defence of the bomber proposals, reiterating once again the rationale behind the Air Ministry's theory of deterrence, and explaining that the heavy bomber was really an economical weapon in comparison to lighter bombers if judged in terms of its capacity to deliver large loads over great distances. In reply Simon reviewed the Treasury's financial objections to the program, then turned to the specifics of the bomber proposals. He pointed out that the heavy bombers would not be in the flying stage of development for at least another two years, hence would hardly act as a deterrent during that time. He suggested that it would not make sense to risk the stability of the economy on such an unknown quantity, and urged his fellow ministers to keep the bomber program under review while authorizing the Air Ministry to place orders for the full number of fighters it had requested. Chamberlain supported the Treasury case, reminding the Cabinet "that it was not by any means certain that we could beat Germany in an armaments race. Such a race was not merely a question of money, but also of industrial capacity and the labour force; in that latter respect Germany had the advantage over us."[23] Noting that four fighters could be built for the cost of one heavy bomber, he urged the Cabinet to support the Treasury's recommendations which they duly did.

The Treasury had hoped that the Cabinet's decision would finally kill the bomber program, but it did nothing of the kind. Within two days of the Cabinet decision Kingsley Wood wrote a strong letter to Chamberlain in which he urged that the bomber

[23] CAB 23/96, 53 (38) 2, 11/7/38.

program be retained. Horace Wilson wrote to Chamberlain expressing his opposition to continuing with the bomber on financial grounds. In summing up, however, he wrote, "It is somewhat difficult for laymen to argue a case of this kind against the experts."[24] This was the problem of technical expertise. Warren Fisher had voiced the same complaint a year earlier, and in the course of 1939 it was to rob the Treasury of most of its control over the Air Ministry program. Despite the Cabinet decision, the full bomber program was included in the estimates for the 1939 budget that the Air Ministry submitted to the Treasury. The ministry insisted that it must be given authorization to place orders if there were to be any bomber program, and supported its contention with the argument that: a. the peak of production of the aircraft was determined by the size and scale of the plant that had to be constructed to produce it; b. makers would not proceed without definite orders; and c. those orders had to be large enough to insure a fifteen-month run on full production to justify the cost of recruiting and training labor and setting up the plant. These were the technical arguments with which the industry was confronting the Air Ministry, and the Air Ministry was in turn forcing them on the Treasury. The Treasury lacked the expertise to break that chain of argument, hence was faced with the choice of either insisting that the bomber program be dropped entirely or acquiescing to the Air Ministry's requests. Simon wrote resignedly, "We must agree. But I would like to consider a personal note to Sir K. W. (with a very few figures) foreboding the inevitable smash if spending goes on increasing."[25] The bomber program drew both the Treasury and the Air Ministry into a costly cycle of technological imperatives that resulted in the Government's having to meet every demand for additional funds pressed on it by the aircraft industry. By July it had become apparent that, because of the complexity of the bombers, there would be a nine-month delay in getting them into production unless the Government provided the funds to double the labor force and the productive capacity. In the course of a note describing discussions between Air Ministry and

---

[24] T 161/40760/03/3, 12/13/38, Wilson to Chamberlain on the bomber program.
[25] T 161/923/40760/03/3, 1/11/39, Simon note on the Hopkins letter concerning the bomber program.

Treasury officials on this subject, B. W. Gilbert, a Treasury under-secretary, wrote:

> We gathered from one or two remarks which Sir Henry Self [an Air Ministry official] let fall, that Sir Kingsley Wood appreciated that in the aircraft industry he had created a Frankenstein [monster] which was out of control.[26]

As war approached, the tyranny of expertise began to take its toll on the institution of Treasury control.

Financing Rearmament, 1939

The officials at the Treasury were overwhelmed by the defence budget for 1939-40. Gilbert, whose responsibility it was to oversee the making of the budget, wrote in November 1938, after receiving the preliminary defence estimates:

> I am not sure that we shall be able to do much unless the political outlook changes, but in any event 1939 is past paying for and 1940 will run to very large figures even if not as large as the S/S's forecast. Grim![27]

Richard Hopkins expressed his concerns in more concrete but no less pessimistic terms.

> It is true that there are a good many people about who have lost all sense of money questions and all regard for the preservation of our economic resources, but soberer minds are bound to be gravely perturbed by speculations of this character which I should expect to have serious reactions on the international exchange, on credit and on trade from now to the Budget and beyond.[28]

Gilbert projected that defence spending over the next three years would be in the neighborhood of £1,400 million, and that the Government would have to borrow £720 million to cover it. Rationing was to all intents and purposes dead, and more importantly so was the idea that underlay it: "to keep our defence efforts

---

[26] T 161/923/40760/03/3, 7/17/39, Gilbert to Barlow concerning the bomber program.

[27] T 161/949/49094/2, 11/12/38, Gilbert note on finance.

[28] T 175/101, 11/8/38, Hopkins note after learning of the R.A.F. estimates.

within the limits imposed by the necessity of maintaining unim-
paired that economic and financial strength which is just as vital in
war as adequate defence."[29] While Hopkins and Alan Barlow, the
newly appointed Joint Second Secretary at the Treasury, wished
to cling to the ration, Gilbert proposed that the Treasury find a
new criterion for limiting defence spending, suggesting that it
withhold sanction from any new proposals that did not make a
major contribution to defence preparedness in the near future. In
Gilbert's eyes the situation was reaching a crisis point:

> Defence expenditure is now at a level which must seriously call
> into question the country's ability to meet it, and a continuance
> at this level may well result in a situation in which the comple-
> tion of our material preparations against attack is frustrated by a
> weakening of our economic stability which renders us incapa-
> ble of standing the strain of war or even of maintaining those
> material defences in peace.[30]

In reality the limitation of defence expenditure was no longer in
the Treasury's hands. Its only recourse was to warn of the impending financial breakup while seeking the means of raising revenue that would best delay it.

The Treasury's basic problem was finding £70 million more
than it raised in 1938 to cover the estimated increases in defence
costs. While acknowledging that the bulk of that amount would
have to borrowed, the Treasury had doubts about the Government's credit and the impact such borrowing would have on the
pound. Phillips in a note to Hopkins wrote:

> The borrowing program outlined is terrific. Whether so much
> of the national savings can be forced into this channel without
> violent dislocations, such as a sharp rise in interest rates, re-
> mains to be seen.

Pointing out that new taxation to cover the cost of new borrowing
was essential, he continued:

> Unless this is done it will be far more difficult to raise the

[29] T 161/905/35164/39, 1/24/39, Gilbert to Simon on the War Office estimates
for 1939.
[30] T 161/904/34610/39, 1/17/39, Gilbert on the service estimates for 1939.

money as the investing world will most certainly expect to be shown that the nation can easily handle the new interest charges. I do not say that the City is *right* to argue that way, but that is the way they do argue and we must humour them if we want their money.[31]

Richard Hopkins succinctly explained the importance of not doing anything that would give the financial community or the public at large the impression that Britain could not bear the burden of financing the rearmament program.

If and when, in addition to the anxieties in men's minds about impending war, it comes to be realised how completely we are snowed under by mounting expenditure, and it comes to be thought that this country is coming to the end of its tether and is prepared to let things slide, the foreign efflux will be swollen by an internal movement away from sterling. The foreigner after all can take away no more than he possesses here, but there is no limit to what our own people might try to send abroad. . . . It seems essential that we should make whatever effort we can to preserve the confidence of our own people in the continuing value of our currency.[32]

The "foreign efflux" to which Hopkins alluded was a reference to the fact that since Munich foreign investors had been taking their money out of Britain to invest it in places where war appeared less imminent. This placed a considerable drain on Britain's reserves of gold, exacerbating a situation that was already quite serious as a consequence of her balance of payments deficit caused by the rearmament program.[33] These were the very same gold reserves that were the foundation of Britain's international credit, which in turn was the basis upon which rested a major portion of the rationale that finance was the fourth arm of the services. The Treasury was concerned that borrowing, to the extent necessary to cover the costs of the defence program, would cause inflation, which would be yet another blow to the stability of the pound, and the one that would very likely touch off the "internal movement

[31] T 161/949/49094/2, 11/4/38, Phillips to Hopkins on the finance of defence.
[32] T 171/341, 1/3/39, Hopkins on the budget.
[33] T 161/949/49094/2, 11/4/38, Phillips to Hopkins on the finance of defence.

away from sterling," to which Hopkins referred. Still, there was no choice but to borrow. After discussing the situation with Norman, the Governor of the Bank, who concurred, the decision was made to ask Parliament for powers to borrow £400 million over the next two years. In late February the decision was announced, and it was received with relief by a City that was deeply concerned about the possibility of another increase in the income tax. Hopkins noted with surprise that, despite the fact such borrowing was bound to result in some inflation, "the City has accepted the borrowing prospects with surprising equinamity [sic]."[34]

While another increase in the income tax had been ruled out, the Treasury was convinced of the validity of Phillips' argument about the need to raise enough new revenue to cover the cost of the increased borrowing. Where that revenue was to come from was the subject of much contention in the Treasury. Only Horace Wilson favored another raise in the income tax, and Simon explained to him the Treasury's view that "its psychological effect might be very damaging."[35] Within the Treasury attention was turned to the surtax and the estate duty, two levies on the rich that had not been raised since the National Government came to power, despite the ongoing search for revenue for defence. While there was some reluctance to impose an added financial burden on those strata of society that formed an important part of the Government's political constituency, it was acknowledged in Hopkins' words that "possibly some upward readjustment—(not necessarily a flat rate increase)—might be judged necessary to sweeten the pill for the masses."[36] These increases were worked out and agreed upon in February, when they were incorporated into the budget bill along with the increases in such indirect taxes as those on cars, tobacco, and sugar, which bore the major strain of the increase in revenue.[37] The Treasury hoped this combination of new borrowing and taxation would permit the economy to survive another year. While the Treasury's program recognized the realities of the situation, it sought to evade rather than confront

[34] T 175/115, 3/7/39, Hopkins on the prospects for borrowing.
[35] T 171/115, 4/6/39, Simon to Wilson.
[36] T 171/341, 1/3/39, Hopkins on Phillips note of 12/15/38.
[37] Sabine, *British Budgets*, pp. 142-143.

them. It was left to Warren Fisher, whose influence had long since waned, to indicate the direction that should have been taken. He pointed out that, if the nation had not been so blind to the menace of Prussianism,

> we conceivably might have been able to amble along in our old and well tried and (for the small minority) very comfortable economic paths. Which means that business could have gone on profiteering out of the occasion of the country's need, and with this example before him the workman would have been stimulated to insistence on some improvement of his rather shabby lot in life.

> But in fact we have done none of these things (except profiteering) and 1938 is either forgotten or read as an augury of hope.[38]

To deal with the situation he proposed a program that would have taxed the rich heavily, cracked down on profiteers, and made luxuries very dear. It was a just and viable program, but not one that recommended itself to a Government still intent on evading the realities of its situation. Although the Treasury had lost any effective control over defence expenditure, having been forced by events to give up the ration and the policy of meeting all recurrent expenses out of revenue that underlay it, like the Government, it chose to carry on as though nothing had changed, ignoring the first signs of economic disintegration that were embodied in the country's increasing loss of gold.

PART II: THE FAILURE OF COOPERATION

Compulsion and the Ministry of Supply

In light of the military deficiencies revealed in the course of the Munich crisis, the question of whether a Ministry of Supply with powers of compulsion should be created was raised once again. Within the Cabinet the plan had two proponents, Ernest Brown, the Secretary of State for Labour, and Leslie Hore-Belisha, the Secretary of State for War. Brown felt, in view of efforts that would be required to accelerate defence production, and the strain those efforts would put on the nation's labor supply, that the Gov-

[38] T 171/341, 1/3/39, Warren Fisher on the economy.

ernment would have to have the power to compel firms to turn their labor, their plant or both over to defence production if the acceleration of the defence programs were going to succeed.[39] Hore-Belisha believed that, if the Army were to be equipped for action on the Continent within a reasonable span of time, a Ministry of Supply was vital to mobilize industry to begin producing the arms and equipment that would be needed. He felt that reliance on the voluntary cooperation of industry to secure those goods would result in delays and costs that would be unacceptable in light of the gravity of the existing situation. Brown and Hore-Belisha arrived at their opinions as the result of the problems their respective departments were facing in the aftermath of Munich. While both of them believed that a Ministry of Supply with compulsory powers was necessary, neither felt strongly enough about it to insist on it as an issue of principle in the Cabinet.

The Government's position on the subject was expressed by Inskip in a paper written for the Cabinet. He contended that neither the international nor the supply situations was serious enough to warrant the dislocations that the imposition of controls would inevitably entail, and he doubted whether they would be politically feasible. After pointing out that the real bottlenecks in production were the result of technical problems that compulsion could not affect, he noted that industry had cooperated fully up to that time, and expressed his conviction that it would continue to cooperate in the acceleration program. It was his opinion that compulsion would only be necessary when cooperation failed.[40]

Industry of course was very interested in preventing the imposition of controls. A number of the nation's leading industrialists expressed their concern about the creation of a Ministry of Supply to Horace Wilson and offered their assistance in assuring the success of cooperation. Wilson wrote Chamberlain:

> the leaders of industry would say that before deciding to have a Ministry of Supply—which, as has been said over and over again, avails nothing in the absence of compulsory powers—any Department which is experiencing difficulty in obtaining supplies should send for the section of industry concerned and

[39] CAB 2/8, 334 (7), 10/20/38, minutes of C.I.D. meeting on labor shortages.
[40] CAB 24/279, CP 234 (38), 10/21/38, Inskip on supply policy.

say what it is that is wanted.Owing to the decline of ordinary production there seems no reason to suppose the demands cannot be met. If, of course, in any particular section there were opposition, some pressure might have to be brought to bear, a process in which we might expect to get the assistance of the leaders of industry who would not want the hand of the Government forced owing to any recalcitrant individuals or individual firms.[41]

Chamberlain agreed with Inskip about the undesirability of creating a Ministry of Supply in peacetime, and was sensitive to the opposition of the business community to it as well. At a Cabinet meeting on October 26, 1938, at which the issue was discussed, he expressed the view that the potential of cooperation for the expansion of defence production had not been fully exploited, and that it would be both premature and counterproductive to institute any measures of compulsion, such as those the creation of a Ministry of Supply would entail, until it had been. He also pointed out that the very threat of the imposition of controls was a potent weapon that would encourage industry to do everything in its power to cooperate.[42] At that Cabinet meeting it was agreed that the time was not ripe for a Ministry of Supply.

Chamberlain did, however, take advantage of the industrialists' offer of assistance in insuring the success of cooperation by creating an industrial panel to advise and assist the Government on supply questions. Made up of some of the nation's leading industrialists, the panel worked to improve communications between industry and government, assisted in clearing up technical difficulties that arose between government departments and their contractors, gave the Government what might be called the enlightened capitalist's view of proposed Government legislation where it affected business, and on occasion brought pressure to bear on firms whose cooperation with the Government left something to be desired.

The Case of the Machine Tool Industry

Inskip's and Chamberlain's assurances to the Cabinet that industry had cooperated fully with the Government on the rearma-

[41] PREM 1/336, 10/21/38, a letter from Wilson to Chamberlain.
[42] CAB 23/96, 50 (38) 4, 10/26/38.

ment program were given with the full knowledge that one of the key industries in the program, the machine tool industry, had willfully refused its cooperation on an issue of importance to the Government. Since early 1937 the machine tool industry, which built the machines that made the parts for aircraft, guns, and every other industrial product, had refused to permit the Government to inspect its books to verify that the costs of some of the products the Government was purchasing from it were reasonable. It was important for the Government to have such access in order to assure both the Parliament and the public that there was no profiteering. All of the industries but the machine tool industry recognized that, if they did not cooperate in permitting costings when they were requested, the Government would have no political alternative but to take powers to obtain them. No Government could permit even the appearance that it was being anything but vigilant in preventing profiteering if it valued its political survival. By June 1938 the Government had concluded that the only way it was going to gain access to the books of machine tool firms was through legislation but, fearing the reaction of the business community, it drew back from such action. When the industrial panel was formed in December, it was asked to try to bring some pressure to bear. The industrialists, who had no desire to see the Government pass legislation that would enable it "to walk into any factory at any time they like to take a costing,"[43] did their best to show the machine tool people that their recalcitrance would only make things worse for everyone, but to no avail. It was due in large part to the machine tool industry's refusal to cooperate that the Government wrote powers to compel industry into its Ministry of Supply Bill in May of 1939.

Aircraft Industry Profiteering

The machine tool industry case, which became a matter of public knowledge in March 1939, is an overt example of the breakdown of cooperation. There arose between the aircraft industry and the Government a situation in which cooperation broke down in a very different, far more significant way. The aircraft industry did not openly defy the Government; it simply took advantage of its economic position to realize excessive profits, thereby violat-

[43] CAB 16/225, IP 41st meeting, 4/26/39.

ing the trust that was supposed to be the touchstone of the doctrine of cooperation. The understanding the Government had reached with industry was that the Government would not take powers to compel industry, and that industry would organize itself to meet the Government's defence requirements at a reasonable cost. That was the basis of the doctrine of cooperation. No laws were passed making profiteering illegal; it was simply understood that such behavior, though economically rational, was unacceptable. The Government had issued repeated assurances that no one would be permitted to realize excessive profits from the nation's need, and had steadfastly maintained in Parliament that it did not require powers of compulsion to insure that those assurances were fulfilled. Unfortunately, with the aircraft industry this was not the case.

It will be recalled that, in the course of the negotiations for the McLintock Agreement, one of the central points of contention between the Air Ministry (which was supported on this point by the Society of British Aircraft Manufacturers) and the Treasury was the method by which the rate of profit was to be figured. The Treasury insisted that "the method of assessing profits most likely to lead directly to equitable results is to assess profits by reference to a fair percentage upon capital actually engaged for the time which it is employed."[44] The Air Ministry, however, insisted that profits be figured on a contract-by-contract turnover basis, pointing out the practical difficulties inherent in calculating how much of whose capital was involved for how long.[45] Although the Treasury forced many revisions of the terms of the agreement, it did not succeed in having the rate of profit made on aircraft contracts related in any way to the amount of private capital invested in the firms executing those contracts. Instead the agreement dealt solely with individual contracts, and the only mention of any rate of profit had to do with the minimum a firm could expect to earn on a given contract (5 percent on turnover). Provisions were also written into the contracts that permitted government accountants to inspect the records concerning the costs involved in executing the contracts—the source of the difficulties with the machine tool

[44] T 161/922/40730/04, 12/16/36, Bridges to Barlow on profits.
[45] See Chapter III, pp. 115-118.

industry. No provision was made, however, for the Government to have access to the trading accounts of firms working on Government contracts, a provision that would have been necessary if the rate of profit on private invested capital were to be calculated.

This lack of access to trading accounts disturbed the officials at the Treasury because it left them with no way of figuring the aircraft industry's rate of profit, and left open the possibility that, despite the Government's public protestations to the contrary, profiteering could go on without their knowledge. They were by no means reassured when they were told by some contractors "that they will hide away their profits," in order to spare the Government embarrassment. In response one of the Treasury's contract specialists wrote:

> This may be good camouflage but it is not good business. Even as camouflage there is a certain problem which merits some investigation: how far a tolerably small firm can in fact hide profits away without the process, (though not necessarily the volume) becoming apparent to say experts of the *Economist* if they choose to conduct a survey of aircraft firms. My impression is that the area of concealment is quite small: anyway it is quite repugnant to our ideas that concealment should be accepted by departments as an implementing of the Government's pledge [that they would permit no profiteering].[46]

By early 1937 the Treasury had set out to ascertain how the experts at *The Economist* would go about finding concealed profits, but with little success. The main problem was figuring out how much private capital was employed without access to the firms' trading accounts. The fact that some firms were owned by larger holding companies made the problem that much more difficult. The Treasury officials' concern with the amount of private capital invested in aircraft firms increased as the Government poured more and more capital into the aircraft industry in 1937 and 1938. They provided interest free loans to subsidize up to 60 percent of the cost of building new plant and provided 90 percent and more of the working capital to firms doing Government work, enabling them to complete an increasingly larger volume of work while

[46] T 161/922/40730/04, 12/10/36, Ismay to Bridges.

providing a smaller proportion of the capital. These provisions meant that, although a firm's rate of profit on the turnover of individual contracts might remain reasonable—and this was the only profit rate the Government accountants had the facilities to verify—the firm was able to generate a greater volume of production in the course of a year. This greater volume increased the absolute amount of its profits and, consequently, its rate of profit on the private capital invested in it. Suppose, for example, that before the Government's decision to accelerate the aircraft program, Firm X had £2 million in private capital invested in it. In the course of a year Firm X completed £1,400,000 worth of work and realized a 10 percent rate of profit on turnover, that is £140,000. This meant that its rate of profit as a percentage of the amount of private capital invested in it was 7 percent. In the course of the acceleration program the Government provided £6 million to the firm for the expansion of production, while the firm raised another £2 million in private capital for the same purpose. This enabled the firm to complete £7 million worth of contracts in the course of a year at the same 10 percent rate of profit. The rate of profit on the firm's private capital, now £4 million, however, jumped to 17.5 percent as a result of the expanded volume of work made possible by the influx of Government capital. This is a simple example of the type of phenomenon the Treasury was concerned about, but lacked the facilities to check up on.

In October 1937 Barlow was informed by the Treasury officials who were trying to figure the aircraft industry's rate of profit on invested capital that in 1936 it was about 10 percent, but they stressed the roughness of that estimate. The following March B. W. Gilbert's note to Hopkins, in which he reckoned that the industry's total excess profits for 1937 were probably only about £750,000, provided an insight into the Government's attitude toward profiteering.

> It is of course vitally important to restrict and, if possible eliminate profiteering, but the reason is not so much financial as political and social in the sense that the good name of Government for efficient administration and for fair distribution of burdens between capital and labour is involved. From the financial point of view the complete elimination of profiteering

would not secure any appreciable abatement of the enormous defence bill of to-day.[47]

Amid the uproar concerning the Air Ministry's conduct of the rearmament program in early 1938, the Opposition again alleged that the aircraft industry was making excessive profits. These allegations, combined with the Treasury's own concern about the matter, persuaded it that a serious effort had to be made to learn the facts of the matter. Therefore, in March a committee was formed within the Treasury to investigate the profits of the aircraft industry. Officially entitled the Committee on Contracts, it was usually referred to as the Murray Committee after its chairman, Evelyn Murray, an accountant. Due to the potentially explosive nature of its area of inquiry, its purpose was a closely kept secret known only to the Prime Minister, Inskip, and a few high Treasury officials.

Meanwhile, to quiet the criticisms of the Government's laxity in preventing profiteering that had been voiced in Parliament, the Estimates Committee of the House of Commons conducted an inquiry into prices and profits. The committee obtained its evidence from testimony given before it in closed session by officials from the three services and the Treasury concerning contracting and pricing procedure. Before the committee's hearings began, Treasury officials met to discuss what information they would offer and how they could present their admittedly inadequate efforts to oversee the profits situation in a way that would make it appear they had the whole problem in hand.[48] Although they were painfully aware that they had no way of knowing whether profiteering was going on or not, they had no desire to raise any doubts in the committee's mind. Such doubts could only bring into question the competence of the department and the Government, and give rise to the kind of political crisis that it is always best to avoid. The other departments were equally reassuring in their testimony, and the committee's final report duly concluded: "The three Defence Departments and the Treasury appear to be fully alive to the importance of keeping a continuous and detailed

[47] T 175/102, 2/16/38, Gilbert to Hopkins on profiteering.
[48] T 161/855/44933, 6/10/38, minutes of a Treasury meeting to determine the line of testimony to be given before the Estimates Committee.

watch upon costs and profits and of improving and strengthening their methods as experience dictates."[49]

Although the Estimates Committee's report succeeded in quelling the general criticism of the Government's contract procedure, one Labour MP, R. R. Stokes, continued to insist that industry was exploiting the Government. A man who had produced artillery shells under Government contract as recently as 1936, he knew from experience what was going on. He wrote to Simon:

> What I have never yet got successfully across to either you or any of the Government Departments is that in the eyes of the business community, the Government's approach to this problem [of contracts for defence material] is regarded as nothing short of a joke. . . .[50]

Shortly after Munich, Gilbert, at the Treasury, learned that Hawker Siddeley Aircraft, a primary government contractor, was going to declare a dividend for 1938 of 32.5 percent on its stock, plus a 10 percent bonus. Realizing that this revelation would immediately revive the questions in Commons about the Government's determination to prevent profiteering, he resolved to learn what kind of profits the aircraft industry was making by any means he could, so that the Treasury could prepare an adequate reply. On October 17 he wrote Gregg, the head of the Inland Revenue (tax) section of the Treasury, explaining the situation and asking him to compute the average profits for aircraft firms in 1937 from the tax records. Gregg was not pleased with the request, pointing out the dubious legality of invading the confidentiality of tax returns, but agreed to comply. In the month and a half that it took Inland Revenue to prepare the figures for Gilbert, many of the aircraft manufacturers presented their annual reports, and while most were not so injudicious as to declare dividends the size of Hawker Siddeley's, the dividends were sufficiently generous to renew the furor over profiteering with a vengeance. Inland Revenue turned its figures over to Gilbert on December 1 with the admonition to keep both the figures and their source strictly secret. The figures were, in Gilbert's words, "to say the least of it,

---

[49] T 161/942/44425, 7/11/38, report of the Estimates Committee.
[50] T 161/950/51936, 7/22/38, Stokes to Simon on profiteering.

disquieting."[51] They revealed that in 1937 the average rate of profit on private invested capital in the aircraft industry was 20 percent, and that the firms that did 75 percent of the Air Ministry's work realized profits of from 21.8-23.5 percent.[52] The figures for 1938 of course were not available, as the tax returns for that year had not yet been filed, but it was clear that they would be even higher due to the great increase in the amount spent on Air Ministry programs. Although the Treasury had studiously avoided arriving at any figure that it felt could be regarded as a justifiable rate of profit, it was clear that the profits being earned by the aircraft industry under the McLintock Agreement were unacceptable, if for no other reason than that they were indefensible from a political point of view.[53] The Treasury sent its information to the Air Ministry with a certain degree of self-satisfaction. Wrote a Treasury official to Gilbert:

> This is extremely interesting and shows up the A. M. badly. What becomes of D[onald] B[ank]'s [the Air Ministry's Director of Contracts] general assurances that the A. M. are only allowing reasonable profits?[54]

It is interesting to note that, outside of informing the Air Ministry and two other Treasury civil servants, Gilbert did nothing more with the information he had gained. Meanwhile, a delegation of MPs from the Conservative Party, led by Stanley Reed, visited Simon on December 16 to ask him about profits in the aircraft industry, and how they should handle constituents' queries on the subject. On January 1, 1939, Reed sent a letter to Simon

[51] T 161/922/40730/04, 12/5/38, note from Gilbert to Gregg thanking him for the information.

[52] T 161/922/40730/04, 12/1/38, Gregg to Gilbert on the profits of the aircraft manufacturers for 1937.

[53] The Treasury's reluctance to set a specific figure as the maximum justifiable rate of profit stemmed in part from the fact that "under present conditions . . . no contractor can be compelled to undertake Government orders: if the store in question is essential, and there is no other contractor to make it, the Government has got to offer him a profit which he will consider worth his while to earn" (T 161/942/44435, 5/14/38, Treasury memorandum by Compton). It was not until the middle of the war that the Government arrived at 7½% as the justifiable rate for low risk work, and 15% as the upper limit for high risk propositions (AVIA 15/38751 Sec Policy 44, 1943).

[54] T 161/922/40730/04, 12/1/38, Compton to Gilbert on the Inland Revenue figures.

containing a set of figures detailing the rates of profit for 1938 on invested private capital in the four leading aircraft manufacturers that showed firms realizing profits as high as 42 percent. This was apparently news to Simon, whose secretary wrote Gilbert that, "He [Simon] was rather hot about this and said he wanted them [the figures] investigated."[55] Why Gilbert hesitated for more than a month before turning his obviously important information over to the ministerial head of the department is difficult to fathom. He may have forgotten it in the pressure of considering the defence estimates, but this is unlikely. More probably he was reluctant to bear the bad news that the department had failed to implement the Government's pledges adequately, and had consequently put it in a potentially disastrous political position. It was not until the middle of January that the Government began to come to grips with the problem.

The Murray Committee, which had received the Inland Revenue figures in mid-December 1938, completed in February of the following year an interim report on aircraft contracts that explained the problem and made recommendations for remedying it. Pointing out that changes in the circumstances under which aircraft were produced had taken place since the guidelines for the McLintock Agreement had been evolved, the committee explained:

> In the first place any marked increase in turnover in relation to capital employed must mean that a profit, calculated as a reasonable percentage of cost, must result in an unreasonable return on such capital. Secondly, the introduction of Government capital assistance relieves the industry, to that extent, of the necessity of meeting the standing charges on capital which they would have otherwise had to provide. Thirdly, with the development of straight-run production the risk of loss is now not greater than a commercial business might be expected to shoulder.

In view of these considerations, the committee continued:

> We have no doubt that the scale of remuneration laid down in the agreement does in fact enable the contractors to make pro-

[55] T 161/922/40730/04, 1/4/39, Simon's secretary to Gilbert.

fits which in many cases are excessive at present and will become more so as expansion proceeds and straight-run production becomes more general. For these reasons we have reached the conclusion that the McLintock Agreement is quite unsuitable to meet the conditions in which the industry is supplying aircraft in the present day.[56]

The report closed by recommending that the agreement be renegotiated to take into consideration the existing realities, and suggesting, if the industry did not agree, that "the Government would have no option but to invite Parliament to confer upon them adequate powers in some form or other to deal with a situation which would become increasingly intolerable."

The Government's Response

The question of how the problem was to be dealt with was an exceedingly difficult one. In the event, the decisions concerning it were made on the administrative level within the Treasury and the Air Ministry. The Cabinet was never informed of the situation, and presumably the Prime Minister, the Chancellor of the Exchequer, and the Secretary of State for Air were the only ministers aware of it.

Initially there was a sharp difference of opinion between the Treasury and the Air Ministry over what action was to be taken to rectify the problem. The Air Ministry pointed out that "the agreement was expressly made for the duration of the expansion programme and contains no provision for variation or termination."[57] It was a legally binding contract between the industry and the Government, and there was no way of forcing the industry to renegotiate it from the standpoint of law. The Air Ministry's initial proposal in light of this difficulty was that the Treasury should create a special tax that would skim the excess profits from all industries including the aircraft industry.[58] It is clear that this proposal was made with the intention of sparing the Air Ministry from once more incurring the ill will of the industry with which it had

[56] T 161/1189/46398, 3/10/39, Murray Committee interim report on contract procedure.

[57] T 161/922/40730/04, 2/20/39, the Air Ministry view of the profits problem. A letter from Arthur Street to Kingsley Wood.

[58] AIR 6/36, 149th meeting of the Committee on Air Expansion, 1/11/39.

to work so closely. Such a tax was the last thing the Treasury wanted to consider. The Treasury officials' memories of the difficulties of enforcing the Excess Profits Duty in the last war and the uproar caused by the proposal of the National Defence Contribution in 1937 combined with their unwillingness to alienate the business community disposed them strongly against the tax. It was soon decided that the only alternative was for Kingsley Wood to confront the S.B.A.C. with the information in the Government's hands and insist that the agreement be renegotiated. As the Murray Committee report noted, the threat of the Government's taking compulsory powers could be used as a means of persuasion. The Government could also threaten the industry with the public exposure of the situation to force them to reconsider. The Government and presumably the industry realized, however, that either of these actions could cut both ways, being as damaging to the Government as to the industry. For this reason the Government fervently hoped that the industry would understand the logic of the situation and submit gracefully.

Until Kingsley Wood made his approach to the industry, the ministers involved decided, no hint of the problem would be made in Parliament. Wood was very concerned

> That if an announcement is made in the House which leads them [the industry] to suppose that the agreement is to be broken or varied, or even that the question of their profits is the subject of a special enquiry, there will be an explosion and not only will the present friendly relations be impaired, but it will become much more difficult to get alterations made by agreement.[59]

On February 23 a heated debate over the method by which profits should be computed on government contracts took place in Parliament. In the course of the debate the Government let it be known that the Murray Committee was investigating the question, but gave no indication of the results of the inquiry. Four days later, during a continuation of the debate, Kingsley Wood defended the Air Ministry's procedure of calculating profit on the basis of

[59] T 161/922/40730/04, 2/20/39, Street to Wood, a summary of a note from Street to Barlow.

turnover and the reasonableness of the profits that procedure permitted. Although he conceded that

> It is perfectly true that, in criticising profits and the methods and the machinery which we have adopted in relation to fixing them, you can make certain calculations, for instance, on the basis of nominal capital, and try by that means to build up a case of profiteering.

(as he knew the Treasury had already done), he questioned the validity of any such case. Knowing, however, that it soon would be necessary to temper some of his more confident statements about the soundness of his ministry's contracting procedure, he did admit that "It may be that further evidence will be forthcoming and that the system may require to be amended in some particular respect."[60] By telling less than he knew, Wood was able to avoid both open prevarication and any mention of the inquiry into the aircraft industry's profits.

On March 1 Kingsley Wood confronted Charles Bruce-Gardiner with the situation as the Government understood it, and presented its proposals to remedy it. Bruce-Gardiner gave little response, stating only that he would have to discuss the matter with the members of the industry. It was more than a month before the two parties met again.

In the interim, the Government felt it was necessary to make some limited concessions concerning the Air Ministry's contracting procedures. Kingsley Wood went before the House of Commons on March 9 to present in general terms the findings of the Murray Committee concerning the McLintock Agreement. While avoiding any mention of excessive profits, he discussed briefly some of the "changes in circumstances which have taken place since the agreement was framed," and allowed that "some modification of its terms is now desirable." He concluded by telling the House that, upon learning of these problems, he "at once got in touch with representatives of the industry, who showed every readiness to meet me and arrive at reasonable conclusions."[61]

[60] *Parliamentary Debates* (House of Commons), Vol. 344, 2/27/39, col. 957-958.
[61] Ibid., col. 2395.

Lacking any specific details, the Opposition was unable to take much advantage of this vague admission that something was amiss. Speaking for Labour, Garro Jones pointed out that the Government had never made public the terms of the McLintock Agreement, which he referred to as "the basis of the racket," and that the Government had refused "to disclose to the House of Commons the formula by which those [aircraft] prices are arrived at."[62] Hugh Dalton reiterated Labour's belief that "gigantic fortunes are being made" and that the Government was "making millionaires nearly as fast as we are making aeroplanes,"[63] but lacking any convincing figures, he was unable to make a viable case for his assertion.

The press was also unable and for the most part unwilling to go any further into the matter. After Kingsley Wood's speech announcing the reopening of discussions with the aircraft industry concerning the terms of the McLintock Agreement, *The Economist* stated that it was "scarcely likely" that there had been, as the Government continued to insist, no undue profits made by the aircraft industry, but did not press the issue further. *The Times*, on the other hand, welcomed Wood's announcement as "outstanding proof of [the] reluctance both on the part of the Government to favour, and on the part of industrialists to practice profiteering."[64]

When the Air Ministry and the S.B.A.C. met in early April to discuss modifications in the terms of the McLintock Agreement, the primary reluctance in evidence was that of the industry to make any changes whatever in the contract arrangement. Sir William McLintock once again represented the industry, and he quickly made it clear that it felt that the existing agreement was perfectly fair to all involved. At the following meeting he "repeated that the percentages proposed in the Air Ministry formula were quite inadequate in the eyes of the industry." He went on to say that "neither did the industry welcome the idea of the present

[62] Ibid., col., 2475.

[63] Ibid., col. 2405.

[64] "Air Progress and Profits," *The Economist*, 3/18/39, pp. 555-556; editorial, *The Times*, 3/10/39, p. 17.

McLintock agreement being scrapped. His own view was that it gave the Air Ministry all they needed."[65]

The talks remained stalemated for two months, with the industry refusing to consider any revision of the agreement. It was only after the Ministry of Supply Bill was passed in May, giving the Government arbitrary powers upon the outbreak of war to review and have submitted to arbitration any contracts they felt permitted excessive profits, that the S.B.A.C. began to reconsider its position. War was clearly imminent, and the industry did not want to find itself in a position in which the Government could and certainly would dictate the terms under which it produced aircraft. Feeling that it would be preferable to negotiate a new contract while it still had some leverage, the S.B.A.C. began to bargain in June. The new terms, which were agreed on in mid-July, were based on the Air Ministry's calculation that the industry had been earning an 8 percent to 9 percent profit on turnover under the old agreement, and that its total profits for 1939 would amount to £6 million. This meant that the annual rate of profit on the estimated £28 million in private capital invested in the industry would be 21 percent.[66] Under the terms of the revised agreement the industry agreed to relinquish a third of those profits, £2 million, and to share equally with the government any further earnings in excess of those realized in 1938. Despite these apparently considerable concessions, Wood warned Simon that the industry would still earn a large aggregate profit due to the sheer size of the Air Ministry program, and explained:

> There are limits below which the profit rate cannot be reduced in relation to turnover. That limit is fixed by considerations of equitable treatment as between one industry and another and I am satisfied that the arrangement now made with the aircraft industry is the best that could be negotiated, having regard to the reasonable claims of both sides.[67]

[65] T 161/1189/46398, 4/3/39 and 4/13/39, discussions with McLintock on a revision of the McLintock Agreement.

[66] AVIA 10/243, 10/15/39, minutes of a meeting at which Sir Harold Howitt, an accountant who had recently been appointed to the Air Council, was briefed on the background of the McLintock Agreement.

[67] T 161/950/59136, 6/1/39, Wood to Simon.

Despite these assurances, which the Treasury accepted, the Government had in fact once again been taken advantage of by McLintock and the aircraft industry. In December 1940 the Chief Accountant of the Directorate of Contracts concluded that this revision of the agreement continued to permit the industry to realize exorbitant profits, and strongly urged that legislation be passed to correct the situation.[68]

The story of the aircraft industry's profiteering, and of the Government's reaction to it, provides some insights into the relationship between the Government and business. That relationship was in many ways defined by the doctrine of cooperation, which the behavior of the aircraft industry did so much to undermine. Basic to that doctrine was the belief shared by the National Government and the business community that what was good for business was good for the nation. The doctrine of cooperation emerged from this common understanding in the form of an agreement between the Government and the business community. In this agreement the Government undertook to carry out the rearmament program without interfering with industry, while industry undertook to organize itself in a manner that would enable it to meet the nation's arms requirements quickly and at reasonable cost. The Government made it clear in both word and deed that it believed reasonable profits were not only justified but were the moving force behind the efficient operation of the economy. However, the question of what constituted a reasonable profit proved to be the point on which the doctrine of cooperation eventually foundered. Implicit in the doctrine was the understanding that business would not exploit opportunities created by the rearmament program to realize windfall profits. Such restraint was expected to result from the realization that such exploitation would be detrimental to the political survival of the Government, which benefited business in so many ways. However, when the opportunity to realize windfall profits presented itself, the nation's leading defence industry never hesitated. Putting its interest in profit before all else, it dealt the Government and the doctrine of cooperation a serious blow.

The Government's refusal even to consider compulsion in the

[68] AVIA 15/1039, 12/30/40, note on profits under the second McLintock Agreement by the chief accountant of the Directorate of Contracts.

face of this obvious failure of the doctrine of cooperation remains to be explained. Although both Chamberlain and Inskip had stated that the Government would take powers if cooperation failed, when the time came, political considerations prevented such action. If the Government had taken powers, its members believed that they would have lost the support of the business community, which was the political base of their party and without which they could not govern effectively. The National Government was, or more importantly believed itself to be, the prisoner of its constituency. When the aircraft industry set its interests above those of the nation, the Government felt it had no alternative but to act as it did, keeping the matter quiet and trying to show the industry the error of its ways. Whether the Government was in fact as dependent on the support of the business community as it believed itself to be is really beside the point. The Government's actions were in accordance with its members' understanding of their position, and as this case demonstrates, they understood themselves to be dependent on the support of the business community. Where the interests of business clashed with the interests of the nation, the National Government did its best to reconcile the latter with the former.

PART III: THE IMMINENCE OF WAR

The Revival of the Ministry of Supply

Little more than two months after he had submitted papers to the Cabinet in which he dismissed the need to create a Ministry of Supply in time of peace, Thomas Inskip reversed his position. In early January 1939 he began to suggest that the deterioration of the international situation made war imminent, and that in view of this the Government should begin to set up the organization that would oversee the conduct of that war in order to facilitate the nation's transition to a war situation. In so doing he brought into focus the discontent of virtually every Cabinet minister involved in defence planning with the defence policy being pursued by the Prime Minister. Although the reasons for his change of position are nowhere stated, they are not difficult to surmise. By the end of December it had become clear that appeasement was a failure, and that it was only a matter of time before Germany went to war.

263

It was also clear that Britain's plans for the mobilization of her resources on the outbreak of war were inadequate, a fact that had been brought home by the Munich crisis and reiterated frequently by Inskip's adviser on supply questions, Arthur Robinson. It was probably Robinson who convinced Inskip that if the social and economic chaos that marked the beginning of the First World War was to be avoided, a Ministry of Supply should be in place at the outbreak of war to coordinate the mobilization of the nation.

Inskip was aware of the shortcomings of the existing plans for mobilization because he was the chairman of the subcommittee on supply organization in war. This committee was responsible for overseeing the efforts of the Central Planning Organization, a joint committee of the Board of Trade and the Principal Supply Officers, which did the actual planning for the conversion of civilian industry to war production. Under the existing scheme a Ministry of Supply was to come into being at the outbreak of war and implement the arrangements previously made by the Central Planning Organization. While this approach appeared to be workable, those involved in executing it knew it was deficient in several important respects. The civilian industries that were to be converted to war production had not been informed of the role they were expected to play because the supply organization considered their plans too tentative to act on. It also had not been determined, even in theory, what powers, if any, the Ministry of Supply was to have to control wages and prices, powers that were vital to the control of a war economy, because the Treasury was loath to deal with the question. Finally, the existing planning organization lacked the manpower and the funding to refine its plans any further. Arthur Robinson did not hesitate to suggest that, if the nation had to go to war under the plans as they were then constituted, chaos would result.

It was in light of this bleak prognosis that Inskip prepared a paper for the C.I.D. recommending that a decision be made to begin to organize the Ministry of Supply in peace. He pointed out that, under existing plans the reserves of the services, in combination with the existing capacity of the armaments industries that supplied them, were supposed to be sufficient to make good the losses in materiel suffered in the first twelve months of war, during which time the Ministry of Supply was expected to have successfully completed the conversion of civilian industries to war production. This, however, was not the case in two important regards. First, the reserves of the Army and the R.A.F., in combination with the capacities of their supporting industries, were insufficient to meet the anticipated attrition of those first twelve months; and second, the plans for the mobilization of the nation's industrial capacity were not adequate to affect the projected conversion of civilian industry in the specified time. If the nation had to fight under such circumstances, it would be seriously endangered by its inability to replace equipment as fast as it was lost. He concluded that, in order to rectify the situation, it was necessary to begin immediately to expand the nation's capacity to produce arms and to take steps to organize a Ministry of Supply.[69]

On January 26, 1939, the C.I.D. met to consider Inskip's proposals. After Inskip opened the meeting by reiterating his key points, Hore-Belisha pointed out that an earlier Cabinet decision—that there should be no more creation of war capacity before the outbreak of war—was predicated on the assumption that the next war would be one of limited liability, which was not the case with the war that was imminent. Ernest Brown, the Secretary of State for Labour, followed by making the point that only 35 percent of the nation's engineering capacity was engaged in defence work, and suggested that a Ministry of Supply, with powers if necessary, was required before the outbreak of war to turn more of that capacity over to defence. After Kingsley Wood voiced his agreement, Oliver Stanley, the President of the Board of Trade, whose ministry was partly responsible for creating the plans for converting industry to a war footing, expressed his belief that only a Ministry of Supply created in peacetime could adequately lay the groundwork for a smooth conversion to war production. Brown then made the point that the key problem was the mobilization of industrial resources, which would be possible without compulsion. He added that, in pressing for a Ministry of Supply, it was important that the issue of compulsion not be allowed to sidetrack it. The group quickly agreed that the im-

[69] CAB 4/29, 1505-B, 1/18/39, Inskip's suggestions for altering the plans for supply organization in war.

mediate creation of a Ministry of Supply without compulsory powers should be the object of their efforts, and Inskip was given the job of presenting the proposal to the Cabinet.[70]

In a paper introducing the subject to the Cabinet, Inskip explained that he was bringing the subject up once more because he and the other members of the C.I.D. had "the feeling that the situation has so far changed for the worse since October that the decision then made does not hold good now." After citing the strong political current in Parliament favoring a Ministry of Supply, he went on to explain that it would take two to three months to get such a ministry set up and functioning. He continued:

> There was also the feeling [at the C.I.D.] that, if we are anyhow to envisage a Ministry of Supply in war, its establishment in war conditions would be so difficult an operation as to make it most advisable that the outbreak of war should find it already established and running as a unit of administration.[71]

He concluded his paper by stating, a little pathetically, "I have not thought it right, for obvious reasons, to express definite opinions on the difficult questions discussed in this memorandum." It was his last communication as Minister for the Co-ordination of Defence.

Inskip Sacked

On January 17, after learning of the paper that Inskip was preparing for the C.I.D., Chamberlain had asked Inskip to resign his position. His only hint of a reason for the dismissal was the comment that "Your position in the H. of C. + in the country + in the press has gone back lately." A moment later he undercut that explanation when he added by way of condolence that he was "sure there wouldn't be any great exaltation at your departure as there would be if Hore-Belisha was to be going."[72] It is apparent that

[70] CAB 2/8, 245 (1), 1/26/39, C.I.D. minutes.

[71] CAB 24/283, CP 33 (39), 1/28/39, Inskip on a Ministry of Supply.

[72] Inskip Diary 2, 1/17/39. Chamberlain was referring to the brief revolt in December of three junior ministers who threatened to resign from the Government in protest against the shortcomings of the rearmament program if the men they held responsible for them, primarily Hore-Belisha but also Inskip, were not replaced. Led by Robert Hudson, the Secretary of the Department of Overseas Trade, the insurrection, which was apparently promoted by members of the Army's General

Chamberlain's actual reason for asking for Inskip's resignation was that the latter's shift to the view that appeasement had failed and that the nation should do all it could to prepare for the inevitable war conflicted with the policy that Chamberlain sought to pursue, and threatened his control of the Cabinet. It is clear from the January 26 proceedings of the C.I.D. that there was agreement among all the ministers whose departments had to deal with defence questions that further preparations had to be made for the defence of the nation. Even Halifax, a member of Chamberlain's inner circle, was constrained to admit that it was necessary "to think on wider lines than a few months before."[73] Only the new First Lord of the Admiralty, Stanhope, did not concur. With the weight of Inskip's expertise behind them, there was every possibility that the ministers could have persuaded the Cabinet to alter the Prime Minister's policy. This is not to suggest that Inskip had political motives for adopting his new position. He clearly did not. He simply acted in accordance with his assessment of the situation. It is apparent from his diary that he did not even really understand why he had been asked to step down. That Chamberlain kept him in the Cabinet indicates that he did not feel that Inskip's action was politically motivated. It is safe to say that Inskip was a victim of his honesty and circumstance. His contributions to the nation's rearmament program were among the most significant made by any member of the National Government.

The new Minister for the Co-ordination of Defence was Lord Chatfield, formerly of the Admiralty. His first task as minister was to prepare a brief with the assistance of the new Chancellor of the Duchy of Lancaster, W. S. Morrison, that would refute Inskip's case for the immediate creation of a Ministry of Supply. In the paper presenting their arguments, they reviewed the importance of a Ministry of Supply in war and conceded that there were certain organizational advantages to be gained from a Ministry of Supply in peace as well, but concluded that it would be best to

---

Staff who still resented Hore-Belisha's efforts at reorganization, received a brief flurry of publicity in the press before Chamberlain brought the three recalcitrants to see the error of their ways. R. J. Minney provides a succinct account of the episode in *The Private Papers of Hore-Belisha*, pp. 161-166.

[73] CAB 2/6, 345 (1), 1/26/39.

7. "Thin Ice"—Hore-Belisha, Inskip, and Halifax take a sprawl after Robert Hudson and his associates threaten to resign over the inadequacy of the rearmament program. *Daily Express*, December 22, 1938.

postpone its creation until the actual outbreak of hostilities. This conclusion was based on two contentions. The first was that such an organization would not be desirable in peacetime because it would break the line of responsibility from supplier to user. Because the Ministry of Supply would become the common purchaser for all the services, it was argued that each service would lose the direct contact it presently had with its suppliers and thus be unable to exercise the same degree of control over the goods it received. The second contention, contrary to Inskip's argument that it would be far more efficient to set up a Ministry of Supply before the outbreak of war, was that setting up such a ministry with a minimum of dislocations after the start of a war would be feasible if adequate planning were done beforehand. In view of these considerations, they concluded, the liabilities of creating such a ministry in peace outweighed the benefits.[74]

In an annex to the Chatfield-Morrison paper Ernest Brown made some telling points for a Ministry of Supply in peacetime. He explained that such an organization could coordinate the distribution of subcontracts far more efficiently than was currently being done, thus ease the pressure for skilled labor, which had reached the point where contractors were poaching labor from one another. This practice of course was driving up both wages and prices, something the Government had sought to avoid. Furthermore, he pointed out that a supply organization would be necessary before war actually broke out to prevent the hoarding of vital materials in the anticipation of it. He noted that hoarders would not wait until war was a fact to begin their activities, and that if the Government expected to control the allocation of such materials during war to prevent shortages in key sectors, it would have to be prepared to anticipate such activity and deal with it when it occurred.

On March 1 the industrialists on Chamberlain's industrial panel wrote a paper in which they expressed their opposition to the creation of a Ministry of Supply in peace, even without powers. Although they had just completed a review of the Government's plans for the mobilization of industry on the outbreak of war and found them wholly inadequate, they opposed the creation of a

[74] CAB 24/283, CP 48 (39), 2/21/39, Chatfield and Morrison on the Ministry of Supply question.

ministry in peace because they believed it would slow down the completion of the rearmament program then in progress. They also opposed its implementation on the outbreak of hostilities because it would add to the inevitable confusion inherent in the changeover from peace to war conditions. In their view it was best to wait until the war situation had stabilized before setting up the ministry in accordance with carefully prearranged plans.[75]

The various papers on the subject were discussed at the Cabinet meeting on March 2, 1939. In the course of the discussion Kingsley Wood reversed his position at the January C.I.D., explaining that the Air Ministry was having no trouble procuring supplies under the existing system, and he saw no reason to upset it. It was left to Hore-Belisha to make the case for the immediate creation of the ministry. He asserted that, if it were accepted that the ministry would be necessary in war, it was logical to set it up in peace, if only to avoid the inevitable dislocations that would occur as a result of trying to set it up under the stress of war. He also explained that the Army needed the ministry to help carry out the recently authorized expansion, which the Army lacked the staff to execute without assistance. Oliver Stanley supported Hore-Belisha, but to no avail. Chamberlain concluded that the needs of the War Office alone were not sufficient to justify the creation of such an organization, and that the public clamor for a Ministry of Supply should be ignored because it was based on the erroneous belief that the Government was having trouble procuring the goods it required under the existing supply arrangement. The issue of the Ministry of Supply was again dismissed.

In reaching this decision the Cabinet had evaded the central issues. The problem that had caused Inskip to raise the question of a ministry in the first place was not whether the Government was currently obtaining the goods it required, but whether, when war broke out, it would be able to set up a Ministry of Supply and complete the conversion of civilian industries to the production of arms quickly enough to prevent the shortages of supplies that had plagued the British services in the First World War, when their reserves ran out long before the nation was able to shift its produc-

[75] CAB 16/228, IP #59, 3/1/39, the industrial panel's view of a Ministry of Supply.

tive capacity to take up the slack. Basic to the urgency felt by the proponents of the Ministry of Supply was their sense that war was imminent, and that only a short time remained for the nation to prepare itself in peace. Oliver Stanley cogently expressed this opinion at the February 2 Cabinet meeting.

> From one point of view we were already at war and had been for some time. He thought that it was contrary to reality to aggregate defence expenditure over a five year period up to March 1942, and to say that we could not afford it. It was clear that some of the conditions under which we were now living could not last much longer—perhaps not for another year—and the present was probably the crucial year.[76]

Neither Chamberlain nor the Treasury was prepared to face that reality until forced to. Their fears about the political and economic ramifications of the acceptance of Stanley's view, entailing as it would have the relaxation of the last restrictions on defence expenditure and the resort to some form of compulsion in peacetime, prevented them from acknowledging its validity until events left them with no choice.

The Fall of Prague

The fall of Prague on March 15, 1939, forced the Government to confront the reality it had been trying so hard to deny. After a momentary hesitation, during which the Government's leaders had tentatively suggested that the elimination of Czechoslovakia had changed nothing, the weight of public opinion drove home the realization that such a position was no longer tenable. Reluctantly, the Government turned to the preparation for war.

The Army was again the greatest beneficiary of the crisis. The Cabinet decided that an expansion of the Army was necessary to shore up France's less than overwhelming resolve to fight Germany, and to impress upon the Germans that Britain was determined to resist further aggression. In early April the Army presented its requests for doubling the size of the Territorial Army, expanding the regular Army, and expanding industrial capacity to meet the requirements of the War Office as quickly as possible. It

---

[76] CAB 23/97, 5 (39) 3, 2/2/39.

was clear that, despite the resolve such efforts would demonstrate, they would not provide any sort of short term deterrent because it would take at least eighteen months to equip the projected new forces, and longer than that to create sufficient industrial capacity to support them.[77] The Treasury accepted the need for such steps, but raised the continuing question of whether the nation could afford to take them. The Treasury was clearly appalled by the cost. Simon wrote to Chamberlain:

> Has anybody worked out what this really involves in additional factories, additional tools, additional labour and additional finance? So far as I can see the desideratum is stated without any real enquiry into how it is to be achieved. . . . Is it possible to maintain a great Fleet, an immense Air Force requiring a vast labour force behind it, to sustain the dislocation of continuous bombardment from the air, to provide munitions at the rate contemplated for allies as well as ourselves, and at the same time fight with an unlimited Army on the Continent backed by an unlimited supply of material?[78]

The Treasury doubted it, pointing out that the financial situation was as serious if not worse than it had been when the Cabinet decided that there was a limit to what the nation could spend without risking economic collapse. The Treasury was to be consistent in its efforts to limit defence spending until the beginning of the war.

Conscription and the Ministry of Supply

Two measures of considerable consequence were passed as a direct result of the decision to expand the Army. The first measure initiated conscription, and the second created a Ministry of Supply. While the French had insisted as early as January that Britain introduce conscription in order to demonstrate her determination to give them substantive ground support in the event of a war with Germany, the actual reason for its introduction was the Army's lack of sufficient personnel to man the nation's anti-aircraft positions continuously. Up to this time the Government had resisted any type of conscription out of concern that Labour would oppose it and turn it into a political issue. In addition Baldwin had

[77] PREM 1/296, 3/28/39, letter from Hore-Belisha to Wilson.
[78] PREM 1/296, 4/17/39, Simon to Chamberlain.

pledged in 1936 that the Government would never impose conscription in time of peace, a pledge that Chamberlain had renewed after taking office. Noting that pledge in the course of the discussion of conscription, Chamberlain suggested that "it was a mockery to call the present conditions 'peace.' "[79] When the conscription measure was introduced, it was accompanied by a promise by Chamberlain that the Government intended to eliminate excessive armaments profits, the purpose of which was to assure labor that in the new defence effort there would be "equality of sacrifice."[80] This assurance failed to conciliate either Labour or the Liberals who were outspoken in their opposition to conscription. It passed into law nonetheless, its major importance being as a gesture to the French. Very few men were actually called before the war broke out, in large part because Britain lacked the industrial capacity to equip them. In all, the decision to conscript brought a great deal more trouble than benefit.

On April 19, forty-eight days after they had presented their case against the creation of a Ministry of Supply in peacetime, Chatfield and Morrison again went before the Cabinet to request authorization to set up just such a ministry. The reason for their reversal was that the Army was unequipped to carry out the massive expansion plans that had been put in motion since the fall of Prague and required the assistance of a large bureaucracy. The ministry that was suggested would take over all Army supply and the procurement of the stores used in common by the three services. The Navy and the Air Force would continue to be responsible for the purchase of their specialized equipment, such as aircraft and ships stores. Under the proposed legislation the Government would be empowered in the event of war to turn the ministry into a full scale Ministry of Supply encompassing all three of the services without further recourse to Parliament. Finally, the legislation gave the Government the power to enforce priority for Gov-

[79] CAB 23/99, 22 (39) 3, 4/24/39.

[80] In order to implement this wholly political pledge, the Treasury was asked to draw up a special tax called the "armaments profits duty." This tax, referred to by one Treasury official as "that abomination," was universally disliked within the Government because it made it more difficult to deal with defence contractors. Passed into law in July, it never went into effect, as it was superseded by an "excess profits duty," which covered all business upon the outbreak of the war.

ernment orders and to compel contractors to submit to costings and binding arbitration on prices. The Government was reluctant to include the last group of powers in its bill, but felt it had no choice in view of the machine tool industry's well-publicized intransigence. Powers were not requested, however, to give the Government new leverage in dealing with the aircraft industry; its intransigence was a private embarrassment that the Government had no desire to make public. The Ministry of Supply bill that the Cabinet finally authorized in May was really the weakest version it could have gotten away with politically. There was strong sentiment in the Commons that favored the taking of much more sweeping powers to control both industry and labor, but the Government resisted increased powers in order not to alienate either group. Despite the obvious failure of cooperation with respect to profits in key armament industries, the Government clung to cooperation as the basis of the rearmament program. To the end it recoiled from the imposition of any form of compulsion, taking up such powers only when forced to do so by political circumstances. In the period from the invasion of Czechoslovakia to the invasion of Poland, the changes made in the Government's rearmament policy were changes in form rather than substance. Each of the apparently significant changes were tempered by the same restraints that had been so important in shaping earlier rearmament policy. The expansion of the Army was held back by concerns about its cost, conscription was limited out of deference to organized labor, and the Ministry of Supply was created with a minimum of powers in the hope that the Government could continue to carry out the rearmament program on the basis of cooperation with business. Although the circumstances had changed, the political and economic concerns that shaped the way in which they were met had not.

The Failure of Finance

During the period following the invasion of Czechoslovakia, the Treasury fought a holding action in an attempt to forestall what it regarded as the imminent collapse of the nation's economy. The decision to resort to large-scale borrowing to finance the expansion of the Army, despite warnings about its economic

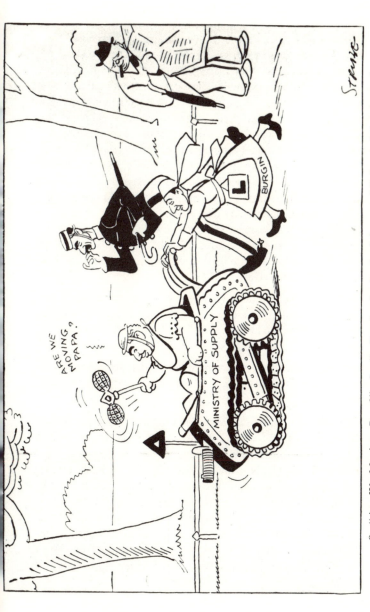

8. "Are We Moving, Papa?"—Leslie Burgin, the head of the new Ministry of Supply, and Chamberlain push Hore-Belisha along as Churchill looks on bemused. *Daily Express*, May 1, 1939.

consequences, was a serious blow, which marked the Government's initial departure from the policy created at the Treasury that had given the nation's financial stability priority over national defence. It was not a departure that the Treasury accepted passively. After acquiescing in the initial surge of plans to shore up the nation's defences, it once again attempted to find a means to order and limit defence spending. The concerns that motivated the Treasury were twofold. First, it doubted that the national savings were sufficient to finance the new Army program without inflation, and second, it was deeply worried about the accelerating deterioration of the nation's international credit position. The two concerns were related, for, if there were inflation, the value of the pound would evaporate and Britain's international credit would evaporate with it. As things were, the outflow of gold had already seriously weakened her credit and showed no signs of abating.

The Treasury's uneasiness about the new direction taken by the Government is clearly conveyed in this passage from a memorandum from Hopkins and Phillips to Simon concerning the proposals to expand the Army:

> we have left behind the time when we were attempting to cut our coat according to the cloth, and so far as finance and the pound are concerned we are sailing upon uncharted waters to an unknown destination. The sea is deceptively smooth.[81]

In mid-May, in the face of new requests for defence expenditure, Simon put the Treasury's view of the issue squarely before the Cabinet.

> We must face hard facts. We cannot finance ourselves by inflationary methods which, if they gave relief for a certain period to an embarrassed Exchequer, would be followed with certainty by a collapse in the purchasing power of our currency, so that the loans we could raise would represent little in buying power. We cannot continue to lose gold in great quantities indefinitely or we shall find ourselves in a position when we should be unable to wage any war other than a brief one. There is a limit to the rate at which we can raise money, and that limit

[81] T 175/104 part #2, 4/15/39, Hopkins and Phillips to Simon on Army expansion.

to the best of my judgement, is already reached. We can go on for another six or nine months if there are no further additions, but thereafter, unless something unexpectedly favourable happens, it may well be that the present rate [of expenditure] could not be maintained.[82]

The hard facts to which Simon referred were these. Britain's yearly national savings were £450 million. The amount already authorized to be borrowed for defence in the current year was £400 million. Under the precepts of classical economic theory a nation cannot borrow more than it has saved without resorting to inflation. In the Treasury's view, borrowing for defence was already perilously close to the level of national savings, and it was concerned that, if additional defence programs were authorized, they would exceed that limit and serious inflation would result.

Maynard Keynes, who was advising the Treasury at that time, prepared a memorandum pointing out that inflation was not inevitable, and that in reality the Government could borrow considerably more than the Treasury believed it could without causing inflation. He explained that in an economy such as Britain's, which was not operating at capacity, when the borrowed funds were spent, they would bring currently idle capacity back into use, thus expanding the nation's production and generating an increase in the national savings. As long as there was idle capacity to be brought back into production, the money that was borrowed, when spent, would increase the nation's savings. Therefore, according to Keynes, Britain, with its millions of unemployed, still had considerable scope for borrowing without having to concern itself about inflation.[83] The Treasury, however, was not convinced by Keynes' argument. In the course of explaining to the Cabinet why the amount the Government could borrow was severely limited, Simon dismissed Keynes' case with the following explanation:

It was argued by Mr. Maynard Keynes and certain other persons that expenditure created savings, and that provided we were prepared to spend money on a large scale, that would au-

[82] CAB 24/287, CP 118 (39), 5/18/39, Simon on finance.
[83] T 175/47, 5/28/39, Keynes memorandum on borrowing.

tomatically create the savings necessary to finance the expenditure. There was, of course, an element of truth in this theory inasmuch as if heavy expenditure increased prosperity, that in turn tended to increase the yield of taxation and national savings. But the fact remained that, although profits which result from increased expenditure would be of assistance if we were now to go in for a large programme of expenditure financed by borrowing, we should, when that programme came to an end, be worse off and not better off.[84]

Had the Treasury heeded Keynes, it would have been spared much of its concern about the size of the rearmament program. As it was, the Treasury embarked on a final effort to keep defence spending within the bounds it felt were imposed by the financial situation. It was an effort that failed to reach fruition by the time war was declared.

The Flight of Gold

While the Treasury worried about the future effect of excessive borrowing, it was plagued by the more immediate, and in the short run more significant, problem of Britain's loss of gold. In the fifteen months following the Anschluss, more than £300 million worth of gold or 40 percent of the nation's holdings had been lost and, despite the imposition of some exchange controls in December and January, the loss continued in 1939. The most important cause of this flight was the concern on the part of foreigners that Britain was about to become involved in war and would freeze their investments in that event. A second important factor was that Britain's balance of trade was in deficit as a result of the fact that 25–30 percent of the raw materials she imported were being used for arms production rather than in production for export. Other factors cited by the Treasury included: a recession in the sterling-bloc nations, which resulted in their withdrawing their gold from London; the fall of the Popular Front in France and the devaluation of the franc, which attracted French gold that had earlier fled the specter of socialism; and, a sizable commitment of

[84] CAB 23/99, 29 (39) 1 and 3, 5/23/39.

what was referred to as "political export credits" to underdeveloped countries.[85]

The reason this efflux was viewed with such great concern was that it seriously undermined what had been referred to in better days as "the fourth arm of the services." Part of Britain's overall war plan was based on the expectation that her strong international credit position would enable her to purchase arms and food abroad through the duration of a long war. Such a use of the nation's international credit had stood her in good stead during the First World War, and it had been planned that her position would be even stronger if another war had to be fought. The loss of gold had come as an unexpected blow to those plans. Alan Barlow of the Treasury explained the state of affairs to a subcommittee of the C.I.D. concerned with strategic planning.

> There had been a fundamental change in our position, and it was clear that we could not regard our financial position as comparable in any way to what it had been before the last war. . . . Everyone knew how shaky our position was, and world opinion was so susceptible that no-one could tell when a slide might not begin. . . . If we were under the impression that we were as well able as in 1914 to conduct a long war we were burying our heads in the sand.[86]

Another consideration that was becoming increasingly ominous was the Johnson Act. Due to the fact that Britain had defaulted on a loan of £850 million made by the United States during the First World War, the Johnson Act had been passed, making it illegal for the American government to lend to Britain.

By early July the gold situation had almost reached crisis proportions. A Treasury memorandum to the Cabinet listed Britain's "War Chest" as consisting of £500 million in gold, which was "rapidly diminishing," and £200 million in foreign securities held in Britain. The memorandum concluded:

> Nevertheless, unless, when the time comes the United States

[85] CAB 24/287, CP 149 (39), 7/3/39, Treasury memorandum on the financial situation.
[86] CAB 16/209, Strategic Affairs subcommittee 4th meeting, 4/6/39.

are prepared either to lend or to give us money as required, the prospects for a long war are becoming exceedingly grim.[87]

At the Cabinet meeting at which this problem was discussed, Richard Hopkins, who was present to explain the technical side of the financial situation, was asked by Kingsley Wood whether the nation's financial position was still sufficiently sound to permit her to conduct a war. Hopkins replied that it was, but that it was becoming weaker every day. Oliver Stanley then noted, "There would, therefore, come a moment which, on a balance of our financial strength and our strength in armaments, was the best time for war to break out."[88] It was a particularly relevant insight. It is possible that Britain's decision to go to war over Poland was based in part on that calculation.

The deterioration of Britain's international credit was in many ways the final summation of the failure of the defence policies that the Treasury did so much to create. Britain's financial strength and her consequent ability to wage a long war formed the basis for the policies of rationing and appeasement. As appeasement failed and war became imminent, rationing was abandoned and gold fled Britain's shores. It is one of the cold ironies of the dismal science that Britain's economic strength, once seen as her greatest asset in preserving peace, should have become one of the liabilities that may have hastened her entry into war.

[87] CAB 24/287, CP 149 (39), 7/3/39, Treasury memorandum on the financial situation.

[88] CAB 23/100, 36 (39) 2, 7/5/39.

CONCLUSION

THE course of British rearmament in the thirties was determined by a complex interplay of forces, interests, personalities, and events. At the center of that interplay was the National Government, which weighed the conflicting considerations
and decided on the response it believed was best suited to the
political, economic, and strategic realities of the time.

The Government's decisions were to a large degree uninflu-
enced by the types of political pressure traditionally associated
with democratic forms of government. The men who guided Britain's rearmament never had to defend their decisions before the
electorate, because the 1935 election, the last before the Second
World War, was held before rearmament had become a major
public concern. The complexity of the issues involved in the
rearmament question, the Government's control and management
of information pertaining to it, and the establishment press's
largely uncritical support of the Government's policy combined to
minimize the possibility of a public outcry against it. Although
there were critics of the rearmament effort on both sides of the
aisle in Parliament, they lacked unity, had little concrete evidence
to support their charges, and were vastly outnumbered by the
Government's supporters. Only on those occasions when events
conspired to assist them were they able to bring pressure to bear
on the Government, and then the Government was able to relieve
that pressure by making token concessions that did not affect the
substance of its policy. The appointment of Inskip as Minister for
the Co-ordination of Defence, in the wake of the criticism of the
Baldwin Government's defence efforts, was one such occasion,
and the sacking of Swinton in 1938, after the attacks on the efficiency of the Air Ministry, was another. Even after the fall of
Prague in 1939 had exposed the bankruptcy of the whole defence
effort, the measures taken by the Government to quiet the public
outcry were less than substantial.

This lack of effective opposition to its defence program from

281

without meant that the course of rearmament was determined by the politics of the decision-making process within the Government. In that process interdepartmental competition, personalities, and, above all, influence played important parts. Departments and interest groups affected by rearmament did what they could to promote policies beneficial to their interests. The policies that were ultimately pursued reflect the distribution of political influence among them.

The conflict between the armed services and the Treasury was the focal point of the debate within the Government over rearmament policy. In this conflict the emergence of first Japan and then Germany as military threats gave the services more than ample material to buttress their cases for expansion. Their lack of unity, however, precluded the possibility of their developing a coherent program and diminished the impact of their proposals to the Cabinet. This enabled the Treasury to forward its own policies by playing the services off against one another. Although the military threat to Britain provided a powerful impetus to rearm, the services were unable to prevail on the Cabinet to insure that the nation's defences were adequate.

The Treasury, on the other hand, was in a much better position to see that its views were heeded. Its responsibility for oversight of all government spending put it at the center of the decision-making process, and its coordination of the Government's efforts to lead Britain out of the economic crisis in the early thirties had given it even more control over the formulation of policy. The Treasury position was further strengthened by the presence of Neville Chamberlain as Chancellor of the Exchequer during the years up to 1937 and as Prime Minister after that. As the second most politically powerful man in the Conservative Party and the heir apparent to Baldwin, Chamberlain had the leverage to insure that his department's positions were carefully considered by the Cabinet. This strong political position combined with the services' lack of unity to enable the Treasury to control the formulation of Britain's rearmament policy.

Although Swinton, Duff Cooper, Hore-Belisha, and Kingsley Wood were forceful advocates of their respective service's programs, their success or failure in winning Cabinet approval of those programs depended on their ability to gain the Treasury's support. Duff Cooper was the most outspoken opponent of Treasury control, but although he was able to make a forceful case for his position, he was seldom supported by the Cabinet. Hore-Belisha also had little success with his pleas for more funds until after Munich, when French pressure forced the Government to reconsider its plans for the Army. Swinton, who was the most successful administrator in the group, owed a large part of his success to the fact that the Treasury felt the Air Force should be the focal point of the rearmament program. Had the Treasury not supported the Air Ministry's program to create the capacity to produce modern aircraft in Britain, Swinton would never have had the opportunity to realize his accomplishment. Kingsley Wood was also an efficient administrator, and adept at squeezing funds from the Treasury. His success in the latter sphere, however, was more attributable to the deteriorating international situation and the burgeoning technological imperatives that made it increasingly difficult for the Treasury to veto Air Ministry projects on their merits than to his political standing. All of these men confronted the Cabinet with the services' views of Britain's strategic situation and their recommendations for improving it, but neither the services nor their heads had the political power to persuade the Cabinet to implement them over Treasury opposition.

On those rare occasions when Treasury opposition was overridden, it was the result of intercession by the Prime Minister, usually on the advice of a close adviser. Advisers who had the trust of the politically powerful, especially of the Prime Minister, enjoyed influence far beyond that inherent in their normal positions. The leading advisers on defence policy in the thirties were Maurice Hankey while Baldwin was in power, and Inskip (until the Munich crisis) during Chamberlain's premiership. It was Hankey who persuaded Baldwin to oppose the Treasury's proposals to unbalance defence of expenditure in favor of the Air Force. Inskip's major contributions were his recommendation, in his final Report on Defence Expenditure in 1938, to maintain defence production at capacity while seeking to appease Britain's enemies, and his intervention on behalf of increased expenditure for the Air Force over Treasury objections in that same year. In both cases he was able to persuade Chamberlain to rule against the wishes of the Treasury, and in both his advice materially furthered the state of Britain's military preparedness. During his tenure as Chancellor of the Exchequer, Chamberlain consulted closely with Warren Fisher on a wide range of issues until their disagreement over the funding of the D.R.C. proposals in 1935, and with Bridges, Hopkins, and Phillips on financial matters throughout. It was the latter three men who were responsible in large part for shaping Chamberlain's financial perspective.

The tenuous and limited nature of an adviser's influence is demonstrated by Fisher's and Inskip's falls from favor. Such influence operates only as long as the advice tendered is in keeping with the views of the patron. An adviser's ability to alter policy extends only as far as his ability to alter the views of the man he is advising. The experiences of Fisher and Inskip testify to the dangers inherent in trying to press such persuasion too far.

In the same way that an adviser retained his influence only as long as his advice inspired the confidence of his patron, a Prime Minister retained his power only as long as his policies and leadership inspired the confidence of his political constituency. In the case of Baldwin and Chamberlain that constituency was the Conservative Party. Although they were the leaders of the party, their power depended on the support of those elements that dominated and controlled it. It is important to keep this in mind when assessing the contributions of both men to the rearmament program. Although Baldwin was responsible for getting the rearmament program under way in 1935, his contribution to its progress after that time was largely negative. Despite the fact that he led the Government to a substantial victory at the polls in November of that year, he was unable to use the momentum generated by the victory to win unified support for his Government's rearmament plans. In large part this failure is attributable to the loss of confidence of important elements of his party in his judgment and capacity to lead following the Hoare-Laval fiasco. This resulted in open criticism of both Baldwin and the Government's rearmament scheme by Conservative backbenchers that eliminated any possibility of galvanizing national support for the program. The decline in Baldwin's health and his lingering premiership further diminished the coherence of defence planning.

Not all the shortcomings of rearmament in its first year are attributable to the effects of the Hoare-Laval affair and Baldwin's failing health, however. Although he felt that rearmament was vital and intended to insure that Britain's defences would not remain deficient, his approach to decision making precluded the possibility of developing a coherent, effective program based on a realistic ordering of national priorities. His strengths as a politician, his ability to conciliate and compromise, were weaknesses when it came to planning. When conflicts over policy arose, he sought to put off final decisions when possible or to work out inefficient compromises when it was not. As a result of Baldwin's shortcomings as an administrator and the unfortunate combination of personal and political circumstances that befell him in 1936, British rearmament proceeded without any real plan or direction until Chamberlain became Prime Minister in 1937. Although many important projects were initiated during that year and a half, Britain would have been in a far stronger military position in August 1939 if the time had been put to better use.

It was left to Chamberlain to bring order to the rearmament program. Upon becoming Prime Minister he ordered a complete review of Britain's responsibilities, the resources she had to meet them, and a reordering of the nation's priorities based on that review that would bring the two more closely into line. The new priorities were then used as the basis for the restructuring of the rearmament program. A unified scheme for the defence of Britain was evolved encompassing economic and foreign policy as well as military planning. Embodied in Inskip's two reports on defence expenditure, this scheme lent coherence for the first time to the rearmament effort. Although the assumptions and assessments on which this scheme was based were faulty in several important respects, resulting in the development of programs that were not adequate to meet the nation's defence requirements, British rearmament was unquestionably made more efficient and effective by the reorganization. It is a testimony to Chamberlain's abilities as an administrator that, despite the restraints he continued to impose

285

on rearmament, Britain was better prepared for war when it came than she would have been had defence planning been allowed to continue as it had under Baldwin.

This is not to say, however, that the Chamberlain government's rearmament program was as effective as it could or should have been. The whole reorganization of the defence program was embarked upon not out of concern for the adequacy of Britain's military preparations, but out of concern about the effect that the increasing cost of those preparations would have on the economy. The Government feared that, if defence expenditure continued to increase, the social and economic stability of the nation would be undermined, and with it, the ability of the nation to defend itself. In the Government's view the stability of the prevailing social and economic order was the ultimate source of Britain's strength. Consequently the preservation of that order was made the primary objective of the reorganization of the defence program. Recognizing that the maintenance of economic prosperity was vital to the realization of its objective, the Government decided to limit defence expenditure to a point where it would not jeopardize that prosperity. Arguing that Britain's financial strength constituted a fourth arm of defence "without which purely military effort would be of no avail," the Government made a conscious choice to take risks with defence rather than with finance. Under its scheme the rationing of defence expenditure and the appeasement of Britain's enemies became the keystones of the Government's plans to insure the security of the nation.

The economic considerations that determined the Government to implement the rationing of defence expenditure stemmed, as has been noted, from its concern about the impact on the economy of the fiscal measures that would be necessary to raise the funds to equip the armed forces to defend Britain adequately. If taxes were increased in an effort to raise that revenue, the result would be a rapid collapse in consumer demand, most probably leading the country back into economic depression. Borrowing, however, also had its dangers, as the Treasury pointed out. Adherents of the doctrines of orthodox finance and classical economics, Treasury officials firmly believed the amount that could be borrowed without causing inflation was strictly limited by the level of national savings. Inflation was undesirable because it would adversely affect trade by undermining the value of the pound, and would give rise to social unrest. Despite the insistence of Keynes and others that far more could be borrowed under present conditions than classical theory indicated, the Government chose not to run the risk of finding out, preferring instead to limit the amounts it borrowed for defence.

The decision to ration defence expenditure made the pursuit of the policy of appeasement necessary. The Government hoped, through appeasement, to diminish the number of Britain's enemies, and thereby decrease the amount it needed to appropriate for defence. At worst the Government believed that through appeasement it could gain time in which to complete rearmament at the slower pace dictated by rationing. In late 1937, when this plan was set forth, the foreign situation made it appear feasible. Germany had committed no belligerent acts since early 1936 when she occupied the Rhineland, a move the Government no longer held against her, and Italy was giving evidence of warming to Chamberlain's personal diplomatic approaches. Although specialists at the Foreign Office continued to insist that the possibility of any kind of substantive détente was remote, and foresaw a worsening of the international situation in the coming year, Chamberlain and the Treasury insisted that matters could be ameliorated if they were approached in the right way. It is indicative of the depth of the Government's concern about the effects of continued increases in defence spending that it was willing to stake Britain's security on what Chamberlain admitted was no more than a possibility. Although Inskip's proposal that defence expenditure during the ensuing two years be as great as necessary to insure maximum production was acted upon, and although the ration was eased in some respects, rationing and appeasement remained the keystones of Britain's defence policy until the fall of Prague in 1939.

While the Government's arguments about the importance of Britain's economic stability were no doubt valid, the risks that it was willing to take with her military preparedness in order to safeguard that stability make it apparent that the Government was as concerned about the social and economic order per se as about

Britain's defences. There was almost an implicit assumption that Britain and her social and economic order were identical, and that if that order were upset, not only could Britain no longer defend herself, but there would no longer be a Britain worth defending. Although this assumption was never articulated, it apparently underlay the Government's contention that Britain's stability was the ultimate source of her strength. It is not likely incidental that the Conservative Party, which dominated the Government, represented the interests of the class that derived the greatest benefit from that order, or that the members of the Government were themselves of that class. Nor is it incidental that the specific restraints placed on rearmament benefitted the elements of the business community that formed the backbone of that class.

Those restraints, which affected the Government's financing of rearmament and its organization of defence production, were imposed for reasons that were political as well as economic. The product of a combination of shared assumptions and subtle pressure, both types of restraint impinged directly on the rate at which rearmament progressed.

The decision to limit defence expenditure, based as it was primarily on the Treasury's concern about the effects of excessive borrowing, had its roots in the economic assumptions shared by the Treasury and the financial community. Embodied in the precepts of classical economics and orthodox finance, those assumptions insured that the City and the Treasury viewed economic questions from the same perspective. The close contact that existed between the two reinforced their shared perspective and guaranteed that the Treasury was aware of the financial community's views on specific issues. It is apparent that the Treasury's whole approach to borrowing was shaped by the City's attitude toward it. This was in large part due to the fact that the success of any Government borrowing operation was dependent on its acceptance by the City. Frederick Phillips stated as much when he wrote:

> I do not say that the City is *right* to argue that way, but that is the way they do argue and we must humour them if we want their money.

In light of this situation there can be little question that the City's opposition to Government borrowing, underscored as it was by its resistance in the marketplace to the defence loan, strongly reinforced the Treasury's reluctance to contemplate more extensive borrowing. Consequently, a measure of the responsibility for the decision to limit defence expenditure should be borne by the financial community.

The industrial community's efforts to persuade the Government not to interfere with the organization of defence production were of necessity more forceful and direct than the subtle methods used by the City to influence financial policy. While the industrialists and the leaders of the Government shared the same assumptions about the undesirability of government interference in the private sector, those assumptions were more open to differences in interpretation than those based on the laws of orthodox finance.

Therefore, in the autumn of 1935 the leaders of industry felt impelled to arrange a secret meeting with Baldwin and Chamberlain to make it clear that they would resist any effort on the Government's part to organize or impose controls on firms involved in the rearmament effort, and urged that it be undertaken on the basis of cooperation between government and industry. Although the Government had never intended to pursue rearmament on any other basis, industry's admonitions served notice as to how strongly it felt about the issue and what reaction the Government could anticipate if it sought to alter its position on the issue of controls.

Rearmament was organized on the basis of cooperation; however, with the essential lubricant, money, in short supply, it proved to be a less than adequate organizing principle. Exacerbated by the shortages of funds, shortages of labor and raw materials led to delays in production. Friction developed between Government ministries and their contractors, and it soon became apparent that, cooperation notwithstanding, their basic interests were in conflict. Whereas the Government was concerned with obtaining as much up-to-date war materiel as possible with the limited funds at its disposal, the contractors were understandably concerned about maintaining the profitability of their undertakings.

289

Profiteering was the final manifestation of the breakdown of cooperation. The product of the laws of supply and demand, it was made inevitable by the Government's unwillingness to intervene in the private sector to insure that the resources at its disposal for national defence were put to optimal use.

The suggestion that compulsion be employed to overcome the problems that bedeviled the rearmament program was first broached to the Cabinet by Swinton in connection with the aircraft industry's difficulties in securing skilled labor. The idea of compulsion was never greeted with much enthusiasm, however, as it was felt that the use of compulsion for such a purpose would antagonize not only the industries from which workers were to be diverted, but organized labor as well. With the Government's belated decision in March 1938 to seek the Trades Union Congress's assistance in dealing with the problem, the question of using compulsion to obtain skilled labor became moot.

At about the same time, however, Swinton, who had become frustrated with what he regarded as the obstructionism of many aircraft firms and generally disillusioned by the Air Ministry's inability to secure through cooperation the aircraft Britain so badly needed, began to advocate the imposition of compulsion or controls in industries that were failing to meet the requirements of the rearmament program. While he was supported on this score by the other two service ministers, Duff Cooper and Hore-Belisha, it was not until after Swinton had been asked to resign that the Munich crisis opened the eyes of a few non-service ministers, like Brown and Stanley, to the peril in which the failure of cooperation had placed Britain.

In the period after Munich the case for the Government's taking limited powers to compel industries involved in the rearmament effort grew more persuasive each month. The inability of defence contractors to overcome through cooperation the delays in the production of weapons vital to Britain's security was sufficient in itself to justify the taking of powers to expedite that production, and the emergence in December of evidence of profiteering simply strengthened that justification by demonstrating that cooperation was not only inefficient, but vulnerable to exploitation.

In light of the failure of cooperation, the Government could have legislated to itself the right to oversee and, if necessary, regulate the prices and profits of industries that produced goods vital to the nation's survival. As the primary supplier of capital to defence industries, the Government could have claimed the rights of the majority shareholder to oversee the management of those industries. Such control would have enabled the Government to expedite the conversion of firms to the production of more modern types of weapons and the implementation of more productive organizational techniques. The creation of a Ministry of Supply with adequate powers would have allowed the Government not only to deal with the waste and delay of existing programs, but to prepare the industry of the nation for the conversion from peacetime to wartime production.

Despite the strength of the case for controls and the efforts of Brown, Stanley, Hore-Belisha, and Inskip to persuade the Cabinet to at least assume the power to impose them, it was not until May 1939 that the force of events produced sufficient political pressure to move the Government to create the Ministry of Supply with its attendant attenuated powers. The opposition in the Government to controls was led throughout by Chamberlain, whose personal conviction as to their inefficiency combined with his concern about industry's opposition to them to convince him that they were undesirable. While it is impossible to ascertain precisely the weight each of these considerations carried in determining Chamberlain's position on the issue, a plausible argument can be made for industry's opposition being the dominant consideration.

The fact that many of the members of the Cabinet who came to see controls as necessary initially had the same reservations about them that Chamberlain did indicates that the case for controls was compelling. His banishment of the two most important proponents of controls, Swinton and later Inskip, from their positions, and his replacement of them with men who shared his views on the issue indicates his concern about the persuasiveness of the case they were making and his doubts about his ability to hold the Cabinet to his opinion. (Swinton and Inskip were the only two important ministers whose resignations he requested during his peacetime premiership.) While it is possible that philosophical conviction alone caused him to oppose the case so many of his ministers found convincing, it is unlikely. An astute politician and a pragmatist to his fingertips, it is doubtful that he would have permitted personal philosophy to deter him from a course of action that was demonstrably beneficial. As he himself said during the speech in which he withdrew his proposal for the N.D.C.,

> I do not think, looking back upon my own record, that I have ever been inclined to show a pig-headed obstinacy. Provided I could get what seemed to me to be the important thing, I have never boggled over particular ways of achieving it nor allowed anything in the way of *amour propre* to prevent my taking what I should call a common sense attitude. . . .[1]

It is more probable that, in his calculation of the costs and benefits of controls, he gave more weight to the likelihood of effective industry resistance than did many of his ministers.

This is not surprising, for he had experienced firsthand the pressure industry could bring to bear while the others had not. Only he and Baldwin had participated in the October 1935 meeting at which the F.B.I. had indicated the nature of industry's response to controls. In 1937 he had a taste of the political pressure that industry could exert when he aroused their ire with the introduction of the National Defence Contribution. Viewing it as a form of compulsion, industry had forced him to withdraw his original proposals and abandon the principle that lay behind them. It was a humiliating defeat, and one he was not likely to forget. Thus in 1938, when sentiment favoring controls began to grow in the Cabinet, he opposed it. The consistent opposition of the panel of industrialists, many of whom had been part of the 1935 delegation and had been active in the campaign against the N.D.C., to any form of controls no doubt strengthened his desire to avoid a confrontation with industry over the issue. As a result of these experiences with industry, when Chamberlain reckoned the advan-

---

[1] *Parliamentary Debates* (House of Commons), Vol. 324, 6/1/37, col. 924.

tages and disadvantages of compulsion, he had ample reason to rate highly the disrupting effect that industry's resistance would have on both rearmament and the economy. This consideration combined with his natural dislike of controls to outweigh the benefits that might have accrued from their implementation.

Had industry been willing to accept limited controls to expedite rearmament, it is likely the Chamberlain would have acceded to the requests for their implementation, and that Britain's defences would have been better prepared as a result. Therefore industry should share with the Government the responsibility for the weaknesses of the rearmament program that would have been ameliorated by the use of controls. While it may be argued that that responsibility should rest wholly with the Prime Minister and his Government, as the decision was ultimately theirs to make, the fact remains that industry's opposition was an important determinant of that decision. Because industry's opposition to controls was expressed consistently and in full knowledge of Britain's military situation, industry must be held accountable for its consequences.

It has been pointed out that had Chamberlain assented to the imposition of controls, industry would not in all probability have reacted as strongly as it had led him to believe it would. This has led to speculation about the course that British rearmament would have followed if the Government had been led by a more independent, dynamic Prime Minister who was willing to run greater risks. If such a leader had decided that a full scale rearmament program involving extensive borrowing and the compulsion of industry was vital to the security of the nation, there certainly were those in both the Cabinet and the Parliament who would have enthusiastically supported him. Although such an initiative would have entailed certain political and economic risks, there is little doubt that, given adequate planning and effort, it could have been carried out by a Prime Minister who was willing to use fully the powers of his office. The primary effort would have had to have been directed toward rallying the public behind the program. If this was effectively done, those in the Government and the business community who opposed full-scale rearmament would have

293

had little choice but to go along. All that was wanting was a Prime Minister with the vision and the courage to provide such leadership.

There is considerable question, however, whether a dynamic, resourceful, independent politician, like the one portrayed above, could have gained the leadership of the dominant political party in the thirties, even if Neville Chamberlain had never existed. Everything we know about the Conservative Party in this period indicates that such a man could not have gained the Conservative leadership. If the party had been disposed toward that kind of leadership, it certainly would have given Churchill more opportunity than it did. The truth is that the party reflected the desires of its constituency, and that constituency wanted leaders who they could be sure would nurture the status quo, men who were reliable and safe. Bonar Law, Baldwin, and Chamberlain were all men of this type, really more managers looking after the interests of the majority share holders than leaders. If there had been no Chamberlain, the man who served in his place would have been from the same mold. The shape and scope of rearmament would have been much the same, although probably less effectively run, for Chamberlain was by far the best administrator on the political scene. The conjuncture of the social, economic, and political realities of the time determined the scope and pace of Britain's rearmament program, and thrust forward the men who carried it out. The limits within which they were free to influence that policy were rather narrow and well defined.

This is not to say that neither the Government nor its leaders should bear any responsibility for the shortcomings of British rearmament, but that they should not bear it alone. Certainly they must be held responsible for staking the national independence on too fine a calculation (Earl Balfour's phrase) for the sake of maintaining the prevailing social and economic order. However the contributions of the business community to that calculation should not be overlooked. Both the city's encouragement of the decision to place financial restraints on rearmament and industry's more active efforts to discourage the implementation of controls affected the Government's decisions in those areas. Consequently, both should share with the Government a measure of the responsibility for the effects of the restraints placed on British rearmament.

Though interesting, the question of responsibility is only one aspect of the larger question of how and why the economic and political considerations that shaped the course of rearmament were accorded the importance they were. The answers to that question provide an insight into the way in which issues vital to the survival of a people are decided in a democratic society.

APPENDIX

Defence Expenditure, 1930-1939
in £

year	Navy	Army	Air Force	total defence expenditure	total government expenditure	defence expenditure as a % of government expenditure
1930	52,274,186	40,243,238	17,631,673	110,149,097	881,036,905	13%
1931	51,041,752	38,623,757	17,868,948	107,534,457	851,117,944	13%
1932	50,164,752	36,137,272	17,057,371	103,359,395	859,310,173	12%
1933	53,443,545	37,540,428	16,700,794	107,684,767	778,231,289	14%
1934	56,616,010	39,691,603	17,607,893	113,915,505	797,067,170	14%
1935	64,887,613	44,654,483	27,515,185	137,057,281	841,834,442	15%
1936	80,976,124	55,015,395	49,995,697	185,987,216	902,193,385	21%
1937	101,892,397	72,675,520	81,799,260	256,367,177	978,980,779	26%
1938	132,437,403	121,542,932	143,499,642	397,479,977	1,033,000,000	38%
1939a	153,666,781	227,261,100	248,061,200	628,989,081	1,453,341,000	43%
1939	181,770,565	242,438,217	294,833,921	719,042,703	1,490,000,000	48%

All figures except those for 1939a represent actual expenditures during the fiscal year beginning March 31, including amounts raised by means of National Defence Loan.

Figures for 1939a represent government estimates of gross expenditure as of July.

The figures are drawn from financial accounts presented to Parliament by the Treasury.

BIBLIOGRAPHY

THIS book is based on primary source material, the preponderance of which is to be found in the Public Record Office in London. Of the Public Record Office's holdings, the Cabinet and Treasury Papers have proved the most relevant to the issues raised in this study.

The Cabinet Papers consist mainly of the minutes of the meetings of various Cabinet committees and subcommittees, and the papers they studied and wrote in formulating and proposing policy. Within the Cabinet Papers the following series proved to be the most useful.

CAB 2-4—The minutes and papers of the Committee of Imperial Defence.

CAB 16—The minutes and papers of interdepartmental committees and subcommittees including the Defence Requirements Sub-Committee (D.R.C.), and the Defence Policy and Requirements Committee (D.P.R.).

CAB 23—The minutes of Cabinet meetings, officially referred to as their conclusions.

CAB 24—The papers submitted for the consideration of the Cabinet, offically the Cabinet papers.

PREM 1—The Prime Minister's miscellaneous papers and correspondence.

The Treasury Papers contain a wide variety of information on virtually every aspect of governmental endeavor. Unfortunately they are indexed in a manner that makes it necessary to use hit and miss techniques and much imagination to locate the material one desires. There is in fact a certain logic to the organization of the Treasury Papers, but it only becomes apparent after many hours of trial and error searching. Despite this shortcoming, they proved to be by far the most valuable source of information on rearmament. The following series were the most important to my research.

T 161—The supply series. This contains the files pertaining to government expenditure. In it are most of the materials having to do with the formulation of the Treasury's policy positions.

T 171—The papers relating to the yearly formulation of the budget.

T 173—The Chancellor of the Exchequer's miscellaneous papers and correspondence.

T 175—The Richard Hopkins papers.

T 177—The Frederick Phillips papers.

The trade associations that permitted me access to their archives included the following.

The Association of British Chambers of Commerce.

The British Steel Corporation, which possesses the archives of the British Iron and Steel Federation.

The Confederation of British Industries, which possesses the archives of the Federation of British Industries, permitted me to see some of the F.B.I.'s publications, although the archives themselves were closed pending their reorganization.

The Society of British Motor Manufacturers.

Private collections to which I was permitted access included the following.

The Baldwin and Templewood (Hoare) collections at the Cambridge University Library.

The Inskip diary and the Hankey, Swinton, and Weir papers at the Churchill College Library, Cambridge.

The Runciman papers at the University of Newcastle-upon-Tyne Library.

I used the following published primary sources.

Great Britain, Parliament, *Parliamentary Debates* (House of Commons), 5th series (1932-39 passim).

The press.

The Daily Herald
The Economist
The Manchester Guardian
The New Statesman
The Times

The memoirs of the important figures of the period, while often pleasurable to read, were not terribly helpful. The following were the most relevant.

Amery, Leopold S. *The Unforgiving Years: 1929-1940.* Vol. III of *My Political Life.* London: Hutchinson, 1955.

Avon, Earl of (Anthony Eden). *Facing the Dictators.* London: Cassell, 1962.

Churchill, Winston S. *The Gathering Storm.* New York: Houghton Mifflin, 1948.

Cooper, Alfred Duff. *Old Men Forget.* London: Rupert Hart-Davis, 1954.

Hore-Belisha, Leslie. *The Private Papers of Hore-Belisha* (edited by R. J. Minney). London: Collins, 1960.

Jones, Thomas. *A Diary with Letters, 1931-1950.* London: Geoffrey Cumberledge, Oxford University Press, 1954.

Liddell Hart, Basil. *Memoirs of a Captain,* Vol. I. London: Cassell, 1965.

Swinton, Earl of (Philip Cunliffe-Lister). *I Remember.* London: Hutchinson, 1948.

————, *Sixty Years of Power* (in collaboration with James Margach). London: Hutchinson, 1966.

Templewood, Viscount (Samuel Hoare). *Nine Troubled Years.* London: Collins, 1954.

Vansittart, Lord (Robert). *The Mist Procession.* London: Hutchinson, 1958.

Secondary sources were consulted primarily to provide background information. They have been divided into five catagories: the British economy in the thirties, the theory of cabinet government, biography, relevant histories of the thirties, and journal articles on the period.

The British Economy in the Thirties

Aldcroft, Derek. *The Inter-war Economy: Britain, 1919-1939.* New York: Columbia University Press, 1970.

Branson, Noreen and Heineman, Margot. *Britain in the 1930's.* New York: Praeger, 1971.

Davies, Ernest. *"National" Capitalism: The Government's Record as a Protector of Private Monopoly.* London: Victor Gollancz, 1939.

Galbraith, John Kenneth. *The Great Crash, 1929*. Boston: Houghton Mifflin, 1955.

Kindleberger, Charles. *The World Depression, 1929-1939*. London: Allen Lane, 1973.

Landes, David S. *The Unbound Prometheus: Technological Change and Industrial Development in Western Europe from 1750 to the Present*. Cambridge: Cambridge University Press, 1969.

Lucas, Arthur F. *Industrial Reconstruction and the Control of Competition: the English Experiment*. New York: Longmans, Green and Co., 1937.

Sabine, B.E.V. *British Budgets in Peace and War: 1932-1945*. London: Allen and Unwin, 1970.

The Theory of Cabinet Government

Beer, Samuel H. *Treasury Control*. Oxford: Clarendon Press, 1957. *British Politics in the Collectivist Age*. New York: Alfred A. Knopf, 1967.

Bridges, Lord (Edward). *The Treasury*. The New Whitehall Series No. 12. London: Allen and Unwin, 1964.

Finer, S. E. *The Anonymous Empire*. London: Pall Mall Press, 1958.

———, "The Political Power of Private Capital." *The Sociological Review*. New Series, Vol. 4, no. 1 (July 1956).

Haxey, Simon. *Tory M.P.* London: Victor Gollancz, 1939.

Jennings, William Ivor. *Cabinet Government*. Cambridge: Cambridge University Press, 1959.

Johnson, Franklyn Arthur. *Defense by Committee: The British Committee of Imperial Defense, 1885-1959*. London and New York: Oxford University Press, 1960.

Mackintosh, John P. *The British Cabinet*. London: Stevens Press, 1962.

Milne, Field Marshal Lord. "The Higher Organisation of National Defence." *The Nineteenth Century After* (March 1936).

Stewart, J. D. *British Pressure Groups*. Oxford: Clarendon Press, 1958.

Biographies

Birkenhead, Earl of. *Halifax*. London: Hamish Hamilton, 1965.

Boyle, Andrew. *Montagu Norman: A Biography*. London: Cassell, 1967.

Clay, Henry. *Lord Norman*. London: Macmillan, 1957.

Colvin, Ian. *Vansittart in Office*. London: Gollancz, 1965.

Feiling, Keith. *The Life of Neville Chamberlain*. London: Lowe and Brydone, 1946.

Harrod, R. F. *The Life of John Maynard Keynes*. New York: Harcourt Brace and Co., 1951.

Macloed, Iain. *Neville Chamberlain*. London: Muller, 1961.

Middlemas, Keith and Barnes, John. *Baldwin: A Biography*. London, Weidenfield and Nicolson, 1969.

Reader, W. J. *Architect of Air Power: The Life of the first Viscount of Eastwood, 1877-1959*. London: Collins, 1968.

Rhodes James, Robert. *Churchill: A Study in Failure, 1900-1939*. New York and Cleveland: The World Publishing Co., 1970.

Roskill, Stephen. *Hankey, Man of Secrets, Volume 3, 1931-1963*. London: Collins , 1974.

Relevant Histories

Ashworth, William. *Contracts and Finance*. History of the Second World War, U.K. Civil Series. London: H.M.S.O., 1953.

Barnett, Correlli. *The Collapse of British Power*. London: Eyre and Methuen, 1972.

Bright, Charles C. "Britain's Search for Security 1930-1936: The Diplomacy of Naval Disarmament and Imperial Defence." Unpublished Ph.D. dissertation, Yale University, 1970.

Clark, Ronald W. *Tizard*. London: Methuen, 1965.

Cole, G.D.H. *A History of the Labour Party from 1914*. London: Routledge and Kegan Paul Ltd.,1948.

Collier, Basil. *The Defence of the United Kingdom*. History of the Second World War, U.K. Military Series. London: H.M.S.O., 1957.

Colvin, Ian. *The Chamberlain Cabinet*. New York: Taplinger Publishing Co., 1971.

Cowling, Maurice. *The Impact of Hitler: British Politics and British Policy, 1933-1940*. London: Cambridge University Press, 1975.

Crozier, W. P. *Off the Record: Political Interviews, 1933-1943* (edited by A.J.P. Taylor). London: Hutchinson, 1973.

Dennis, Peter. *Decision by Default: Peacetime Conscription and British Defence, 1919-1939*. Durham, N.C.: Duke University Press, 1972.

George, Margret. *The Warped Vision: British Foreign Policy, 1933-1939*. Pittsburgh: Pittsburgh University Press, 1965.

Gilbert, Martin. *The Roots of Appeasement*. London: Weidenfield and Nicolson, 1966.

————, and Gott, Richard. *The Appeasers*. Boston: Houghton Mifflin, 1963.

Gordon, Michael R. *Conflict and Consensus in Labour's Foreign Policy, 1914-1965*. Stanford, California: Stanford University Press, 1969.

Hancock, W. K. and Gowing, M. M. *The British War Economy*. History of the Second World War, U.K. Civil Series. London: H.M.S.O., 1949.

Higham, Robin. *Armed Forces in Peacetime: Britain 1918-1940, a Case Study*. London: G. T. Foulis and Co. Ltd., 1962.

————, *Military Intellectuals in Britain, 1918-1940*. New Brunswick, N.J.: Rutgers University Press, 1966.

Middlemas, Keith. *The Diplomacy of Illusion: The British Government and Germany, 1937-1939*. London: Weidenfield and Nicolson, 1972.

Mowat, Charles Loch. *Britain Between the Wars: 1918-1940*. Chicago: The University of Chicago Press, 1955.

Naylor, John F. *Labour's International Policy: The Labour Party in the 1930's*. Boston: Houghton Mifflin, 1969.

Postan, M. M. *British War Production.* History of the Second World War, U.K. Civil Series. London: H.M.S.O., 1952.

Robbins, Keith. *Munich.* London: Cassell, 1968.

Rowse, A. L. *Appeasement: A Study in Political Decline, 1933-1939.* New York: Norton, 1961.

Scott, J. D. and Hughes, Richard. *The Administration of War Production.* History of the Second World War, U.K. Civil Series. London: H.M.S.O., 1955.

Skidelsky, Robert. *Politicians and the Slump: The Labour Government of 1929-1931.* London: Macmillan, 1967.

Taylor, A.J.P. *English History: 1914-1945.* The Oxford History of England, Vol. XV (Sir George Clark, editor). Oxford: Clarendon Press, 1965.

————, *The Origins of the Second World War.* London: Hamish Hamilton, 1961.

Thompson, Neville. *The Anti-Appeasers: Conservative Opposition to Appeasement in the Thirties.* Oxford: Clarendon Press, 1971.

Watt, D. C. *Personalities and Policies.* London: Longmans, 1965.

Webster, Charles and Frankland, Noble. *The Strategic Air Offensive Against Germany, 1939-1945, Vol. I.* History of the Second World War, U.K. Military Series. London: H.M.S.O., 1961.

Wendt, Berndt Jürgen. *Economic Appeasement: Handel und Finanz in der britischen Deutschland-Politik, 1933-1939.* Düsseldorf: Bertelsmann Universitätsverlag, 1971.

Wheeler-Bennett, Sir John. *Munich: Prologue to Tragedy.* New York: Viking Press, 1948.

Journal Articles

Ceadel, Martin. "Interpreting East Fulham." *By Elections in British Politics.* Edited by Chris Cook and John Ramsden. New York: St. Martins, 1973.

Eatwell, Roger. "Munich, Public Opinion and Popular Front." *The Journal of Contemporary History.* Vol. 6, no. 4 (July 1971).

Fearon, Peter. "The British Airframe Industry and the State." *Economic History Review*. Second Series, Vol. 27, no. 2 (May 1974).

Goldman, Aaron. "Sir Robert Vansittart's Search for Italian Co-operation Against Hitler." *The Journal of Contemporary History*. Vol. 9, no. 3 (July 1974).

Heller, R. "East Fulham Revisited." *The Journal of Contemporary History*. Vol. 6, no. 3 (July 1971).

MacDonald, C. A. "Economic Appeasement and the German 'Moderates.' " *Past and Present*. Vol. 56 (August 1972).

Manne, Robert. "The British Decision for Alliance with Russia, May 1939." *The Journal of Contemporary History*. Vol. 9, no. 3 (July 1974).

Parker, R.A.C. "Great Britain, France and the Ethiopian Crisis, 1935-1936." *English Historical Review*. No. 89 (April 1974).

Robertson, James C. "The Hoare-Laval Plan." *The Journal of Contemporary History*. Vol. 10, no. 3 (July 1974).

Stannage, C. T. "The East Fulham by Election, 25 October 1933." *Historical Journal*. Vol. 14, no. 1 (1971).

INDEX

Addison, Dr. Christopher, MP-Labour, 106-108

Admiralty, stymied by Treasury, 43-44; fears Italian hostility, 61; needs French ports, 63; opposes Ministry of Defence, 71; presses for two-power standard, 82-83, 142, 188, 198; on Inskip's review, 173; attacks rationing, 200-203

Air Ministry, plans for home defence, 33, 38, 39; acceleration in 1934, 45-46; estimates of German strength, 50; Scheme C authorized, 51-52; Swinton and Weir join, 53-54; the "shadow scheme," 109-12; and the aircraft industry, 113; on profits, 114-17; negotiations with S.B.A.C., 118-25; Scheme H, 140-42; new parity criteria, 141; bomber strategy challenged, 170-73, 238-42; Scheme J, 184; Scheme K, 188, 199; management criticized, 111, 204-205, 217-19; opposes Treasury ration, 211-16, 222-24; Scheme L, 211. *See also* R.A.F.

Amery, Leopold, MP-Conservative, 86

appeasement, and rearmament, 173-83; Munich and, 224-28; and rationing, 287-88

Army, after W.W.I, 19-20; unpreparedness in 1933, 28; continental role, 33, 35-36, 39, 79-81, 184, 235; estimates cut, 43; debate over role, 136-40; effect of Inskip's review, 164, 169; Hore-Belisha's purge, 169-70; seeks more funds, 188-89, 235; budget reduced, 200; expanded after Prague, 271-72; conscription, 238, 272-73

Attlee, Clement, MP-Labour, 49, 59, 217

Austin, Sir Herbert, auto manufacturer, 111-12

Austria, 197

automobile industry, 109-12

Baldwin, Stanley, Lord President, 15; D.C.M. (32) member, 35; on bombing, 37; air parity pledge, 38; opposes Treasury on D.R.C. report, 40; favors balanced defences, 41, 79; on R.A.F. expansion, 42-43; becomes P.M., 53; 1935 election, 56-60; supports League against Italy, 62, 64; defends Hoare, 66-67; weakness of his leadership, 60, 67, 69-71, 86-87; on the field force and T.A., 81; failing health, 87; and rearmament, 87-89, 284-85; discussion with industry, 95-97; describes cooperation, 98; pledges no profiteering, 103; resigns, 157

Balfour, Sir Arthur, 92

Ball, Joseph, 180

Bank of England, 1931 crisis, 13; role as government's banker, 146; and the National Defence Loan, 157

Banks, Donald, Air Ministry Director of Contracts, 124, 255

Barlow, J.A.N., Treasury Second Secretary, on Germany, 178; need for ration, 243; on aircraft profits, 252; describes gold loss, 279-80

Barrington-Ward, Robert, assistant editor of *The Times*, 68

Beharrell, Sir George, industrialist, 95

Bevin, Ernest, MP-Labour, 128, 217-18

Board of Trade, favors accommodating Italy, 61; role in mobilization, 264-65

Boothby, Robert, MP-Conservative, 148n, 153-55

Bridges, Edward, Treasury Under-Secretary, influence with Chamberlain, 74; on effects of borrowing, 160; on response to Inskip's interim

307

Library of Congress Cataloging in Publication Data

Shay, Robert Paul, Jr. 1947-
  British rearmament in the thirties:
  politics and profits.

  Bibliography: p.
  Includes index.
  1. Great Britain—Defenses.   2. Great Britain—
Politics and government—20th century.   3. Great
Britain—Foreign relations—20th century.   I.   Title.
UA647.S37   355.03'3041   76-45911
ISBN 0-691-05248-4